Autodesk Revit 2022 Architecture Certification Exam Study Guide

Elise Moss

PUBLICATIONS

SDC Publications
P.O. Box 1334
Mission, KS 66222
913-262-2664
www.SDCpublications.com
Publisher: Stephen Schroff

ISBN-13: 978-1-63057-432-1
ISBN-10: 1-63057-432-5

Printed and bound in the United States of America.

Preface

This book is geared towards users who have been using Revit for at least six months and are ready to pursue their official Autodesk certification. You can locate the closest testing center on Autodesk's website. Check with your local reseller about special Exam Days when certification exams can be taken at a discount. Autodesk has restructured their certification exams so they can be taken in your home if you have the proper equipment (web cam and microphone). This edition of the textbook has been restructured as well to align with the new certification exams.

I wrote this book because I taught a certification preparation class at SFSU. I also teach an introductory Revit class using my Revit Basics text. I heard many complaints from students who had taken my Revit Basics class when they had to switch to the Autodesk AOTC courseware for the exam preparation class. Students preferred the step by step easily accessible instruction I use in my texts.

This textbook includes exercises which simulate the knowledge users should have in order to pass the certification exam. I advise my students to do each exercise two or three times to ensure that they understand the user interface and can perform the task with ease.

I have endeavored to make this text as easy to understand and as error-free as possible…however, errors may be present. Please feel free to email me if you have any problems with any of the exercises or questions about Revit in general.

I have posted videos for some of the lessons from this text on YouTube – search for Moss Designs and you will find any videos for my book on that channel.

Exercise files can be accessed and downloaded from the publisher's website at:
www.SDCpublications.com/downloads/978-1-63057-432-1

I have included two sample practice exams, one for the User Certification exam and one of the Professional certification exam. These may only be downloaded using a special access code printed on the inside cover of the textbook. The practice exams are free and do not require any special software, but they are meant to be used on a computer to simulate the same environment as taking the test at an Autodesk testing center.

Please feel free to email me if you have any questions or problems with accessing files or regarding any of the exercises in this text.

Acknowledgements

A special thanks to Rick Rundell, Gary Hercules, Christie Landry, and Steve Burri, as well as numerous Autodesk employees who are tasked with supporting and promoting Revit.

Additional thanks to Gerry Ramsey, Will Harris, Scott Davis, James Cowan, James Balding, Rob Starz, and all the other Revit users out there who provided me with valuable insights into the way they use Revit.

Thanks to Zach Werner for the cover artwork for this textbook along with his technical assistance. Thanks to Karla Werner for her help with editing and formatting to ensure this textbook looks proper for our readers.

My eternal gratitude to my life partner, Ari, my biggest cheerleader throughout our years together.

Elise Moss
Elise_moss@mossdesigns.com

Table of Contents

Lesson Seven
Revit Utilities

Introduction
FAQs on Getting Certified in Revit

The first day of class students are understandably nervous and they have a lot of questions about getting certification. Throughout the class, I am peppered with the same or similar questions.

Certification is done through a website. You can learn more at https://home.pearsonvue.com/autodesk. The website currently has exams for Revit Architecture, Revit Electrical, Revit Mechanical and Revit Structure. You can be certified as a User and as Professional. You no longer need to take the User exam before you take the Professional exam. You can jump straight to the Professional exam, if you like.

In the past, both exams required you to use Revit during the exam to walk through problems. The exam is now entirely – 100% – browser-based. All the questions are multiple choice, fill in the blank, true/false, or point and click. You do not have any other software open beside the browser. This means you have to be really familiar with Revit's dialog boxes and options as you are relying on your memory for most of the exam.

You now have the option to take the exam at a Pearson Vue testing center or in the comfort of your own home. You can go on-line to determine the location of the testing center closest to your location and schedule the exam.

If you opt to take the exam in your home, you still need to "schedule" the exam as a proctor will be monitoring you remotely during the exam. In order to take the exam at home, you need a laptop, workstation or tablet device connected to the internet equipped with a microphone and a webcam. Prior to the exam, you will be required to test your system to ensure the microphone and webcam are functioning properly. You also need to upload your identification and take a "selfie" so the proctor can verify that you are the one taking the exam.

When I took the exam at home, I had to clear off my desk, turn off and unplug all monitors except for the laptop I was using for the exam. The proctor made me pick up my laptop and rotate the laptop 360 degrees so she could see the area I was taking the exam and verify that I had no papers on my desk and nothing that could help me during the exam. You are not allowed to take notes or refer to any reference material during the exam. The only device you are allowed to use is the device you are using to take the exam. Your cell phone is to be stored away from your work area.

The User exam requires about 150 hours of hands-on experience. The Professional certification requires about 1,200 hours or two years of experience with the software.

The Autodesk Certified Professional Exam is 46 questions and 120 minutes. This means you have an average of two and a half minutes per question. If you have a slow internet connection, there may be a lag and this will slow you down. So, make sure you have a good internet connection prior to scheduling an exam outside a testing center.

The Autodesk Certified User Exam was not available yet when I wrote this textbook. They had the subject matters which will be on the exam, but no exams. Normally, I try to take each exam at least twice so I have a good idea of the type of questions being asked and can include similar questions in my textbook and as part of the sample exams.

I have indicated in the exercises which exercises are similar to questions which might appear in the User exam and which might appear in the Professional exam, so you can focus on those exercises which pertain to the exam you plan to take.

What are the exams like?

The first time you take the exam, you will create an account with a login. Be sure to write down your login name and password. You will need this regardless of whether you pass or fail.

If you fail and decide to retake the exam, you want to be able to log in to your account.

If you pass, you want to be able to log in to download your certificate and other data.

Autodesk certification tests use a "secure" browser, which means that you cannot cut and paste or copy from the browser. You cannot take screenshots of the browser.

If you go on Autodesk's site, there is a list of topics covered for each exam. Expect questions on the user interface, navigation, and zooming as well as how to place doors and windows. The topics include collaboration, creating Revit families, linking and monitoring files, as well as more complex wall families.

Exams are timed. This means you only have one to three minutes for each question. You have the ability to "mark" a question to go back if you are unsure. This is a good idea because a question that comes later on in the exam may give you a clue or an idea on how to answer a question you weren't sure about.

Exams pull from a question "bank" and no two exams are exactly alike. Two students sitting next to each other taking the same exam will have entirely different experiences and an entirely different set of questions. However, each exam covers specific topics. For example, you will get at least one question about family parameters. You will probably not get the same question as your neighbor.

At the end of the exam, the browser will display a screen listing the question numbers and indicate any marked or incomplete questions. Any questions where you forgot to select an answer will be marked incomplete. Questions that you marked will display as answered. You can click on those answers and the browser will link you back directly to those questions so you can review them and modify your answers.

I advise my students to mark any questions where they are struggling and move forward, then use any remaining time to review those questions. A student could easily spend ten to fifteen minutes pondering a single question and lose valuable time on the exam.

Once you have completed your review, you will receive a prompt to END the exam. Some students find this confusing as they think they are quitting the exam and not receiving a score. Once you end the exam, you may not change any of your answers. There will be a brief pause and then you will see a screen where you will be notified whether you passed or failed. You will also see your score in each section. Again, you will not see any of the actual questions. However, you will see the topic, so you might see that you scored poorly on Documentation, but you won't know which questions you missed.

Results	100	200	300	400	500	600	700	800	900	1000
Required Score										
Your Score										

Section Analysis			Final Score	
Modeling and Materials		82%	Required Score	700
Families		100%	Your Score	691
Documentation		70%		
Views		71%	Outcome	
Revit Project Management		82%	Fail	X

If you failed the test, you do want to note in which categories you scored poorly. Review those topics both in this guide and in the software to help prepare you to retake the exam. You can sign into the Pearson Vue website and download the report of how you did on the exam and use that as a study guide.

How many times can I take the test?

You can take any test up to three times in a 12-month period. There is a 24 hour waiting period between re-takes, so if you fail the exam on Tuesday, you can come back on Wednesday and try again. Of course, this depends on the testing center where you take the exam. If you do not pass the exam on the second re-take, you must wait five days before retaking the exam again. There is no limit on the number of re-takes. Some testing centers may provide a free voucher for a re-take of the exam. You should check the policy of your testing center and ask if they will provide a free voucher for a re-take in case you fail.

Do a lot of students have to re-take the test or do they pass on the first try?

About half of my students pass the exam on the first try, but it definitely depends on the student. Some people are better at tests than others. My youngest son excels at "multiple guess" style exams. He can pretty much ace any multiple guess exam you give him regardless of the topic. Most students are not so fortunate. Some students find a timed test extremely stressful. For this reason, I have created a simulated version of the exam for my students simply so they can practice taking an online timed exam. This has the effect of "conditioning" their responses, so they are less stressed taking the actual exam.

Some students find the multiple-choice style extremely confusing, especially if they are non-native English speakers. There are exams available in many languages, so if you are not a native English speaker, check with the testing center about the availability of an exam in your native language.

Why take a certification exam?

The competition for jobs is steep and employers can afford to be picky. Being certified provides employers with a sense of security knowing that you passed a difficult exam that requires a basic skill set. It is important to note that the certification exam does not test your ability as a designer or drafter. The certification exam tests your knowledge of the Revit software. This is a fine distinction, but it is an important one.

If you pass the exam, you have the option of having a badge displayed on your LinkedIn profile verifying that you are certified in the Revit software. This may help you convince a recruiter or prospective employer to grant you an interview.

How long is the certification good for?

Your certification is good for a particular year of software. It does not expire. If you pass and are certified for Revit 2022, you are always certified for Revit 2022. That said, you are not certified for Revit 2025 or Revit 2030. Most employers want you to be certified within a couple of years of the most current release, so if you wish to maintain your "competitive edge" in the employment pool, expect that you will have to take the certification exam every two to three years. I recommend students take an "update" class from an Autodesk Authorized Training Center before they take the exam to improve their chances of passing. Autodesk is constantly tweaking and changing the exam format. Check with your testing center about the current requirements for certification. Depending on the testing center, you are always going to test on the current release.

How much does it cost?

It costs $150 USD to take the Professional exam. User exams were not available when I wrote this textbook, so they may be less.

Do I need to be able to use the software to pass the exam?

This sounds like a worse question than intended. Some of my students have taken the Revit classes, but have not actually gotten a job using Revit yet. They are in that Catch-22 situation where an employer requires experience or certification to hire them, but they can't pass the exam because they aren't using the software every day. For those students, I advise some self-discipline where they schedule at least six hours a week for a month where they use the software – even if it is on a "dummy" project – before they take the exam. That will boost the odds in their favor.

What happens if I pass?

Because you are taking the test in a testing center, you want to be sure you wrote down your login information. That way when you get back to your home or office, you can log in to the certification center website and download your certificate. You also can download a logo which shows you are a Certified User or Certified Professional that you can post on your website or print on your business card. Autodesk now offers a badge which you can add to your LinkedIn profile.

How many times can I log into Autodesk's testing center?

You can log in as often as you like. The tests you have taken will be listed as well as whether you passed or failed.

Can everybody see that I failed the test?

Autodesk is kind enough to keep it a secret if you failed the exam. Nobody knows unless you tell them. If you passed the test, people only see that information if you selected the option to post that result. Regardless, only you and Pearson VUE will know whether you passed or failed unless *you* choose to share that information.

What if I need to go to the bathroom during the exam or take a break?

You are allowed to "pause" the test. This will stop the clock. You then alert the proctor that you need to leave the room for a break. When you return, you will need the proctor to approve you to re-enter the test and start the clock again. Even if you take the exam at home, you are monitored during the exam. Just open a chat window and tell the proctor you need to pause the test for a bathroom break.

How many questions can I miss?

A passing score is 700. Because the test is constantly changing, the number of questions can change and the amount of points each question is worth can also vary.

How much time do I have for the exam?

The Professional exam is 120 minutes. I don't know how long the user exam will be.

Can I ask for more time?

You can make arrangements for more time if you are a non-English speaker or have problems with tests. Be sure to speak with the proctor about your concerns. Most proctors will provide more time if you truly need it. However, my experience has been that most students are able to complete the exam with time to spare. I have only had one or two students that felt they "ran out of time."

What happens if the computer crashes during the test?

Don't worry. Your answers will be saved, and the clock will be stopped. Simply reboot your system. Let the proctor know when you are ready to start the exam again, so you can re-enter the testing area in your browser.

Can I have access to the practice exams you set up for your students?

I am including a practice exam for both the User and Professional versions in the exercise files with this text.

What sort of questions do you get in the exams?

Many students complain that the questions are all about Revit software and not about building design or the uniform building code. Keep in mind that this test is to determine your knowledge about Revit software. This is not an exam to see if you are a good architect or designer.

The exams have several question types:

One best answer – this is a multiple-choice style question. You can usually arrive at the best answer by figuring out which answers do NOT apply.

Select all that apply – this can be a confusing question for some users because unless they know how *many* possible correct answers there are, they aren't sure. The test tells you to select 2/3/4 correct answers out of 5 and will prompt you if you select too many or not enough.

Point and click – this has a java-style interface. You will be presented with a picture, and then asked to pick a location on the picture to simulate a user selection. When you pick, a mark will be left on the image to indicate your selection. Each time you pick in a different area, the mark will shift to the new location. You do not have to pick an exact point…a general target area is all that is required.

Matching format – you are probably familiar with this style of question from elementary school. You will be presented with two columns. One column may have assorted terms and the second column the definitions. You then are expected to drag the terms to the correct definition to match the items.

True/False – you will be provided a set of sentences and asked to determine which ones are true or false.

Any tips?

I suggest you read every question at least twice. Some of the wording on the questions is tricky.

Be well rested and be sure to eat before the exam. Most testing centers do not allow food, but they may allow water. Keep in mind that you can take a break if you need one. If you are taking the test at home, you can have water on your desk or work area, but that is it.

Relax. Maintain perspective. This is a test. It is not fatal. If you fail, you will not be the first person to have failed this exam. Failing does not mean you are a bad designer or architect or even a bad person. It just means you need to study the software more.

Remember to write down your login name and password for your account. The proctor will not be able to help you if you forget.

Practice Exams

I have created a user practice exam and a professional practice exam. Do not memorize the answers. The questions on the practice exams will not be the same as the questions on the actual exam, but they may be similar.

Creating and Modifying Components

This lesson addresses the following certification exam questions:

- Create and Modify Grids
- Create and Modify Levels
- Create and Modify Walls
- Create compound walls
- Create a stacked wall

Users should be able to understand the difference between a hosted and non-hosted component. A hosted component is a component that must be placed or constrained to another element. For example, a door or window is hosted by a wall. You should be able to identify what components can be hosted by which elements. Walls are non-hosted. Whether or not a component is hosted is defined by the template used for creating the component. A wall, floor, ceiling or face can be a host.

Some components are level-based, such as furniture, site components, plumbing fixtures, casework, roofs and walls. When you insert a level-based component, it is constrained to that level and can only be moved within that infinite plane.

Components must be loaded into a project before they can be placed. Users can pre-load components into a template, so that they are available in every project.

Users should be familiar with how to use Element and Type Properties of components in order to locate and modify information.

There are three kinds of families in Revit Architecture:
- system families
- loadable families
- in-place families

System families are walls, ceilings, stairs, floors, etc. These are families that can only be created by using an existing family, duplicating, and redefining. These families are loaded into a project using a project template.

Loadable families are external files. These include doors, windows, furniture, and plants.

In-place families are components that are created inside of a project and are unique to that project.

Create and Modify Grids

Grids are system families. Grid lines are finite planes. Grids can be straight lines, arcs, or multi-segmented. In the User exam, you can expect at least one question about grids. It will probably be True/False.

Revit automatically numbers each grid. To change the grid number, click the number, enter the new value, and Click ENTER. You can use letters for grid line values. If you change the first grid number to a letter, all subsequent grid lines update appropriately. Each grid ID must be unique. If you have already assigned an ID, it cannot be used on another grid.

As you draw grid lines, the heads and tails of the lines can align to one another. If grid lines are aligned and you select a line, a lock appears to indicate the alignment. If you move the grid extents, all aligned grid lines move with it.

Grids are Annotation elements. But, unlike most annotation elements, they DO appear across different views. For example, you can draw a grid on your ground floor plan, and it would then appear on the subsequent floors (levels) of your model. You can control the display of grids on different levels using a scope box. Grids are datum elements.

A grid line consists of two main parts: the grid line itself and the Grid Header (i.e. the bubble at the end of the grid line). The default setting is for the grid line to have a grid header at one end only.

On the Professional exam, expect a grid question relating to how to control the view display of grids using 2D extents, 3D extents and/or scope boxes.

Exercise 1-1

AUTODESK.
Certified User

Create and Modify Grids

Drawing Name: **grids.rvt**
Estimated Time to Completion: 20 Minutes

Scope
Create and Modify Grids

Solution

1. Views (all)
    Floor Plans
    Level 1
    Level 2
    Site

    Activate the **Level 1** floor plan.

2. Grid

    Select the **Grid** tool from the Architecture tab on the ribbon.

3.

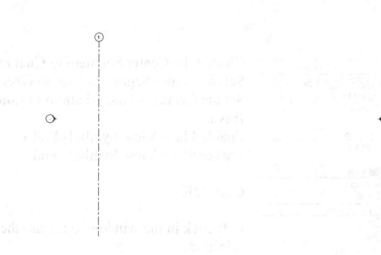

Draw a grid on the left side of the display.

Left pick on the bottom of the screen to start the grid line.

Move the mouse up.

Left pick to place the end point of the grid line.

Cancel out of the command.

Notice that by default the grid line only displays a bubble on one end.

4.　　Select the grid that was just placed.

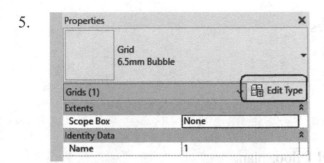

A small box appears at the end of the grid line.

□　　Left click inside the box.

This enables the visibility of the grid bubble.

5.

Properties	✕
Grid 6.5mm Bubble	▼
Grids (1)	🔲 Edit Type
Extents	⌃
Scope Box	None
Identity Data	⌃
Name	1

Select the grid line.
On the Properties panel, select **Edit Type**.

6.

Type Parameters

Parameter	Value
Graphics	
Symbol	M_Grid Head - Circle
Center Segment	Continuous
End Segment Weight	1
End Segment Color	Black
End Segment Pattern	Grid Line
Plan View Symbols End 1 (Default)	☐
Plan View Symbols End 2 (Default)	☑
Non-Plan View Symbols (Default)	Top

Study the properties that are controlled by the grid type.

7.

Type Parameters

Parameter	Value
Graphics	
Symbol	M_Grid Head - Circle
Center Segment	Custom
Center Segment Weight	1
Center Segment Color	Green
Center Segment Pattern	Long Dash
End Segment Weight	1
End Segment Color	Black
End Segment Pattern	Grid Line
End Segments Length	0.0250
Plan View Symbols End 1 (Default)	☑
Plan View Symbols End 2 (Default)	☑
Non-Plan View Symbols (Default)	Top

Change the Center Segment to **Custom**.
Set the Center Segment Color to **Green**.
Set the Center Segment Pattern to **Long Dash.**
Enable Plane View Symbols End 1.
Enable Plane View Symbols End 2.

Click **OK**.

Left click in the window to release the selection.

Notice how the appearance of the grid line changes.

8. 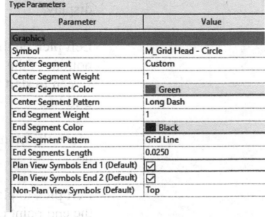 Select the **Grid** tool from the Architecture tab on the ribbon.

9.

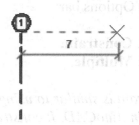

Start a second grid line 7 m to the right of the first grid line.

Use Object Tracking and the Temporary dimension to help you locate the grid line.

Notice that the second grid line has two bubbles visible.

This is because you changed the type properties of the grid.

Select the second grid line.

10.   Select the **Copy** tool from the Modify tab on the ribbon.

11.  On the Options bar:

Enable **Constrain.**
Enable **Multiple**.

Constrain is similar to using ORTHO mode in AutoCAD. It constrains movement in the horizontal/vertical direction.

Multiple allows you to place more than one copy.

12. Select the endpoint below the top bubble on the second grid line.

13. Place four more grids to the right of the existing grids. Don't worry about the dimensions yet.

14. Select the **Aligned Dimension** tool from the Quick Access toolbar located at the top of the window.

15. Select each grid, starting with grid 1.

Then left click above the grids to place the dimension.

This creates a multi-segmented dimension, also known as a dimension string.

16. Left click on the EQ symbol displayed above the dimension.

This toggles the dimension string to set the distance between the grids as equal. This is considered a constraint in Revit.

Click ESC to exit the dimension command or right click and select CANCEL.

17.

Linear Dimension Style
Diagonal - 2.5mm Arial

Dimensions (1)		Edit Type
Graphics		
Leader	☑	
Baseline Offset	0.0000 mm	
Other		
Label	\<None\>	
Equality Display	Equality Text	

Value
Equality Text
Equality Formula

Select the multi-segmented dimension.

On the Properties palette, you can set the Equality Display to Value, Equality Text, or Equality Formula.

Select **Value** from the drop-down.

The dimension now displays a numerical value.

18.

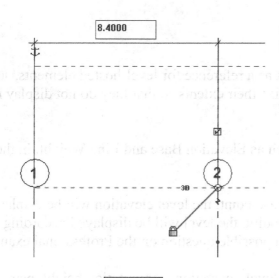

8.4000

Select Grid 2.

Select the dimension value located between grid 1 and grid 2.

Your value may be different from mine depending on how you placed your grids.

19.

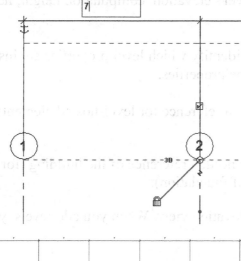

7

Change the value to **7.**

Left click anywhere in the display to release the selection.

20.

Notice that all the dimensions update as they have been set equal.

21.

1 : 100

Click on the **Reveal Constraints** tool located on the display bar at the bottom of the window.

22. The dimensions are displayed as red and bold.

This is because they were defined using an EQ (equal) constraint and are set to always be equal.

23. Click on the Reveal Constraints tool located on the display bar at the bottom of the window to toggle the display off.

24. Save as *ex1-1.rvt*.

Levels

Levels are finite horizontal planes that act as a reference for level-hosted elements, such as roofs, floors, and ceilings. You can resize their extents so that they do not display in certain views.

You can modify level type properties, such as Elevation Base and Line Weight, in the Type Properties dialog.

If the elevation base is set as the project base point, the level elevation will be displayed to the Origin 0,0. If it is set as the survey point, the level will be displayed according to the defined relative coordinates. (This is a possible question on the Professional exam.)

Modify instance properties to specify the level's elevation, computation height, name, and more.

On the certification exam, you may need to identify which level properties are instance properties and which level properties are type properties.

Levels are finite horizontal planes that act as a reference for level-hosted elements, such as roofs, floors, and ceilings.

Create a level for each known story or other needed reference of the building (for example: first floor, top of wall, or bottom of foundation).

To add levels, you must be in a section or elevation view. When you add levels, you can create an associated plan view.

You can resize the extents of a level so that they do not display in certain views.

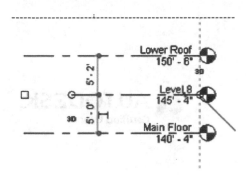

Levels which are blue are story levels. They have floor and ceiling plan views associated to the levels. Levels displayed in black are reference levels and have no associated views.

You should be able to identify the different level components and properties. There will be one question regarding levels in the User exam and in the Professional exam.

You also should be able to identify the components of a level element.

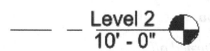

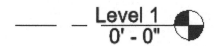

For example, the elevation dimension shown is a permanent dimension when the level is not selected because it is always displayed as long as the level is displayed. If the level is selected, the dimension displayed is a temporary dimension and can be modified.

Exercise 1-2

Create and Modify Levels

Drawing Name: i_levels.rvt
Estimated Time to Completion: 5 Minutes

Scope
Placing a level.

Solution

1.
   ```
   ....... T.O. Parapet
   Elevations (Building Elevation)
       ....... East
       ....... North
       ....... South
       ....... West
   ```
   Activate the **South Elevation**.

   The level names have been turned off.

2. Select each level and place a check in the square that appears. This will turn on visibility of the level name.

3. You should be able to identify the names for each level.

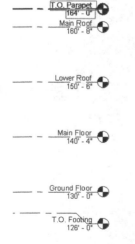

4. 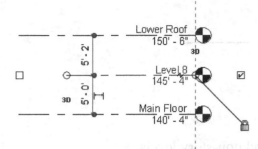 Select the **Level** tool from the Architecture ribbon.

5. 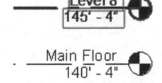 Place a level **5'-0"** above the Main Floor.

6. Note the elevation value for the new level.

7. Close without saving.

Exercise 1-3

Story vs. Non-Story Levels

AUTODESK.
Certified User

Drawing Name: **story_levels.rvt**
Estimated Time to Completion: 15 Minutes

Scope
Understanding the difference between story and non-story levels
Converting a non-story level to a story level

Solution

1. Activate the **South Elevation**.

2. Select each level and place a check in the square that appears. This will turn on visibility of the level name.

3. Study the Main Floor level. Notice that it is the color black while all the other levels are blue.

   The Main Floor level is a non-story or reference level. It does not have a view associated with it.

4. Activate the Architecture tab on the ribbon.
   Select the **Level** tool on the Datum panel.

5. On the Options bar: Uncheck **Make Plan View**. Set the Offset to **8' 0"**.

6. Select the **Pick** tool on the Draw panel.

7. 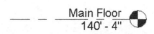 Select the Main Floor level.
Verify that the preview shows the level will be placed 8' 0" ABOVE the Main Floor level.

8. 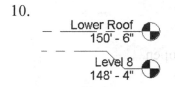 The level is placed above the Main Floor.
Right click and select Cancel twice to exit the Level command.

9. 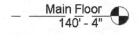 Select the Elbow control on the new level to add a jog.

10. Note that the new level is also a non-story or reference level.
Check in the Project Browser and you will see that no views were created with the new level.

11. 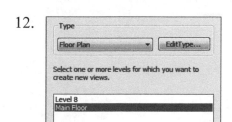 Activate the View ribbon.
Select the **Plan Views→Floor Plan** tool on the Create panel.

12. The reference levels are listed.
Select the **Main Floor** level and click **OK**.

13. The Main Floor floor plan view will open.
Note that the Main Floor floor plan is now listed in the Project Browser; however, there is no ceiling plan for the Main Floor.

- Floor Plans
  - Ground Floor
  - Lower Roof
  - **Main Floor**
  - Main Roof
  - Site
  - T.O. Footing
  - T.O. Parapet

14. 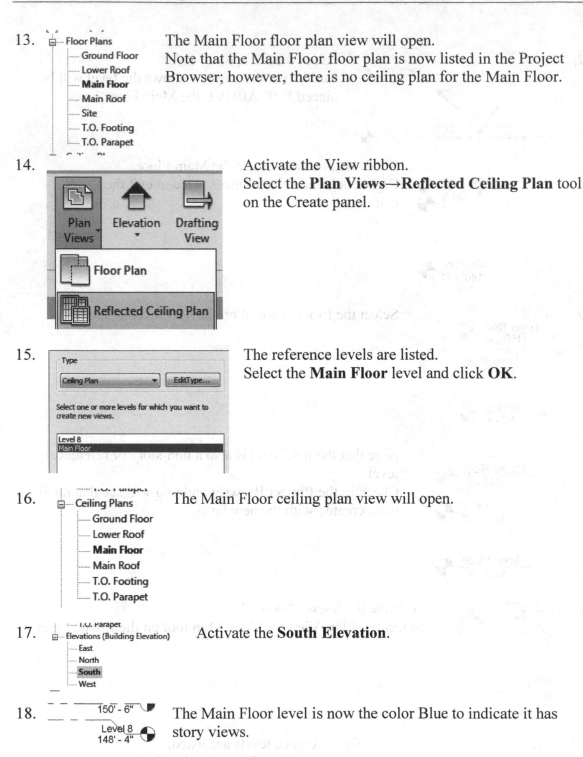 Activate the View ribbon.
Select the **Plan Views→Reflected Ceiling Plan** tool on the Create panel.

15. The reference levels are listed.
Select the **Main Floor** level and click **OK**.

16. The Main Floor ceiling plan view will open.

- Ceiling Plans
  - Ground Floor
  - Lower Roof
  - **Main Floor**
  - Main Roof
  - T.O. Footing
  - T.O. Parapet

17. Activate the **South Elevation**.

- T.O. Parapet
- Elevations (Building Elevation)
  - East
  - North
  - South
  - West

18. The Main Floor level is now the color Blue to indicate it has story views.

150' - 6"

Level 8
148' - 4"

Main Floor
140' - 4"

19. Close the file without saving.

Level and Grid Extents

All levels and grids (also known as datums) have 3D and 2D extents. Datums are Revit elements used as references while modeling. They are finite representations of infinite planes (vertical – grids, horizontal – levels) that are displayed as lines. They are not model elements. They are considered a special category, listed under annotation elements. Unlike other annotation elements, they are not view-specific.

If the 3D extents of the grids don't cross the entire elevation/section/3D view, they will not be displayed.

There are three functions or commands which allow you to control the extents of levels or grids:

- Maximize 3D Extents
- Reset 3D Extents
- Propagate Extents

You need to be familiar with all three of these commands when taking the Professional exam. All three functions are available on the right click menu after selecting the grid/level/section line.

The Maximize 3D Extents function expands the extents of the grid/level/section to the full boundaries of your model.

You cannot reset to 3D extents for a level if crop view is enabled and the level endpoints are not inside the crop area. To use the Reset function, toggle crop view off, use the Reset to 3D Extents and then enable crop view.

The Propagate Extents tool pushes any modifications you apply to a datum object from one view to other parallel views of your choosing. This tool does not work well on levels because the parallel views are essentially mirrored views of each other. For example, the orientation of the South elevation is the opposite of the North elevation; therefore, if you make a change to the extents at the right end of a level in the South elevation, those changes would be propagated to the left end in the North elevation.

The best way to apply the Propagate Extents tool is with the 2D extents of grids. Why only the 2D extents? Because changing the 3D extents affects the datum object throughout the project, independent of any specific view. The 2D extents controls the display of the datum line while 3D extents controls whether the line will appear in other referring views. When you need to adjust the level line in a specific view, but not in the entire model, the level should be set to 2D extents.

Exercise 1-4
Level and Grid Extents

Drawing Name: **datum_extents.rvt**
Estimated Time to Completion: 30 Minutes

Scope
Use Maximize 3D Extents
Use 2D Extents
Use Propagate Extents

Solution

1. 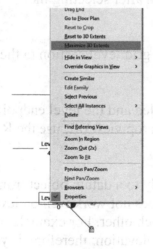   Activate the **South Elevation**.

2.   Select Level 1.

   Right click and select **Maximize 3D Extents**.

3.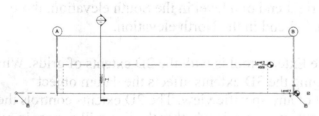

   Notice how the level extends past the building.

   Notice how the bubble at the end of Level 1 is unfilled and it indicates 3D.

4. 🏠 ▾   Switch to a 3D view.

5.
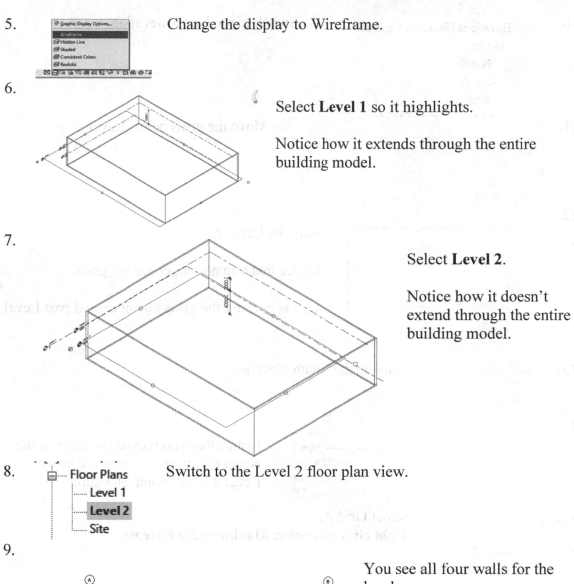

Change the display to Wireframe.

6.
Select **Level 1** so it highlights.

Notice how it extends through the entire building model.

7.
Select **Level 2**.

Notice how it doesn't extend through the entire building model.

8.
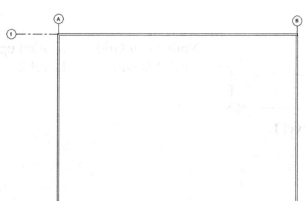
Switch to the Level 2 floor plan view.

9.
You see all four walls for the level.

Floor plan views or parallel views display the entire level as long as they are not cropped regardless of how far the level line extends.

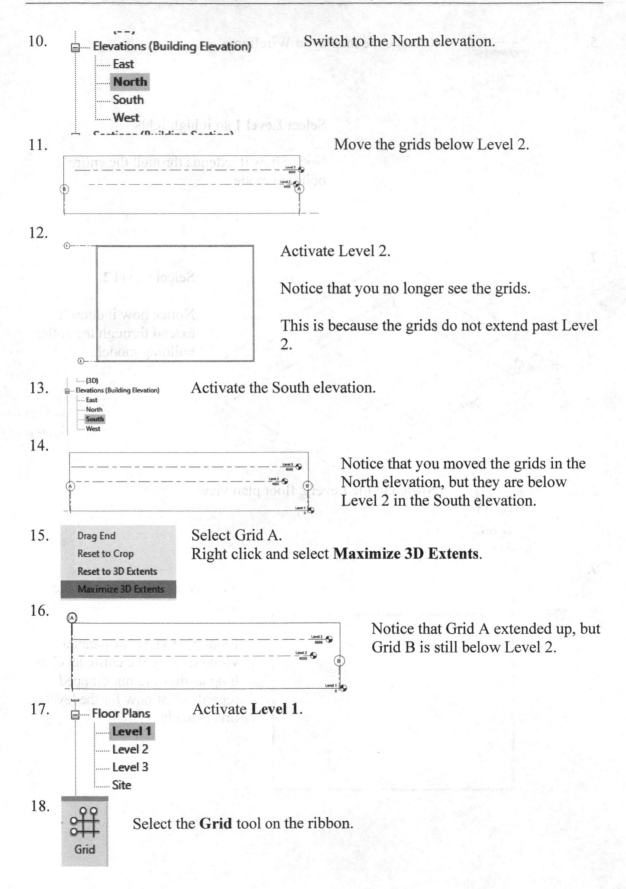

10. Switch to the North elevation.

11. Move the grids below Level 2.

12. Activate Level 2.

Notice that you no longer see the grids.

This is because the grids do not extend past Level 2.

13. Activate the South elevation.

14. Notice that you moved the grids in the North elevation, but they are below Level 2 in the South elevation.

15. Select Grid A.
Right click and select **Maximize 3D Extents**.

16. Notice that Grid A extended up, but Grid B is still below Level 2.

17. Activate **Level 1**.

18. Select the **Grid** tool on the ribbon.

19. 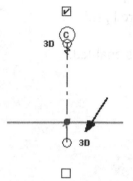 Draw a grid line inside the room.

Do not extend beyond the walls.

20. 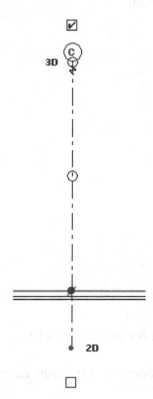 Select the grid that you just placed.

Click the text displayed as 3D to toggle the grid to use 2D extents.

21. Extend the grid below the south wall in the Level 1 floor plan.

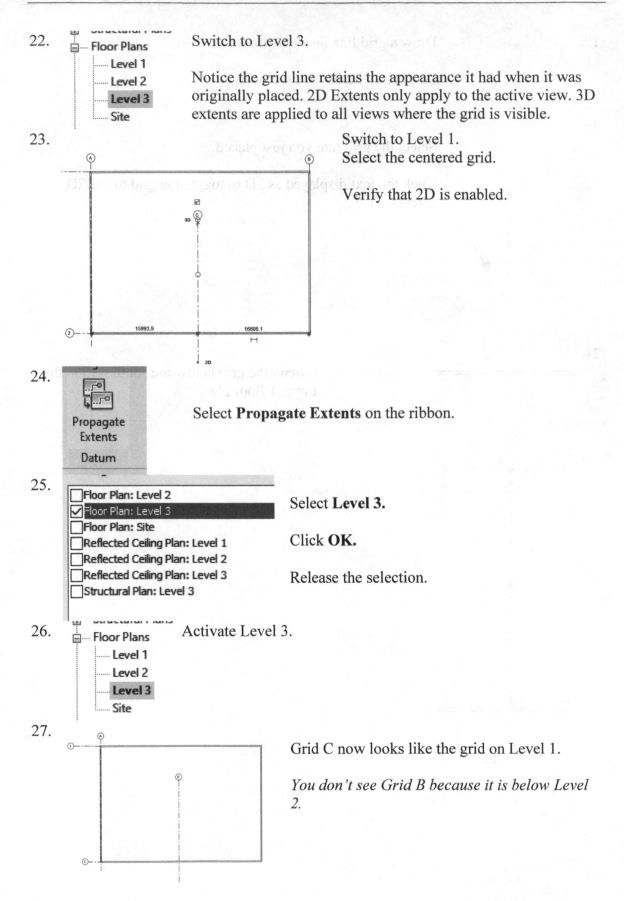

22. Switch to Level 3.

Notice the grid line retains the appearance it had when it was originally placed. 2D Extents only apply to the active view. 3D extents are applied to all views where the grid is visible.

23. Switch to Level 1.
Select the centered grid.

Verify that 2D is enabled.

24. Select **Propagate Extents** on the ribbon.

25. Select **Level 3.**

Click **OK.**

Release the selection.

26. Activate Level 3.

27. Grid C now looks like the grid on Level 1.

You don't see Grid B because it is below Level 2.

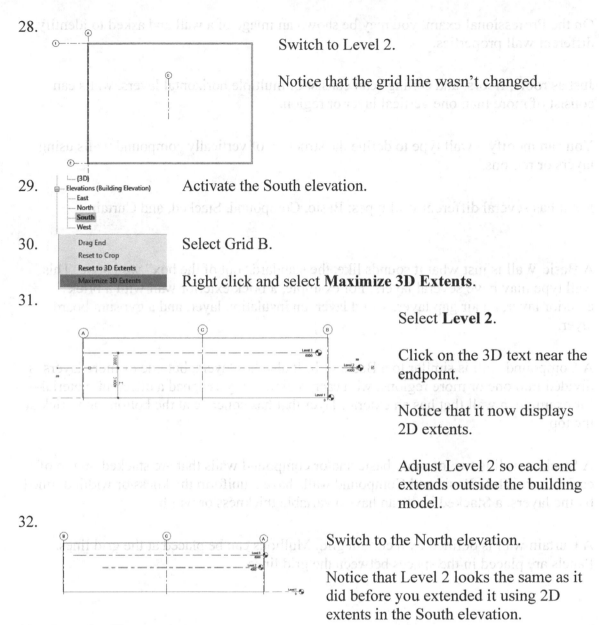

28.   Switch to Level 2.

Notice that the grid line wasn't changed.

29.   Activate the South elevation.

30.   Select Grid B.

Right click and select **Maximize 3D Extents**.

31.   Select **Level 2**.

Click on the 3D text near the endpoint.

Notice that it now displays 2D extents.

Adjust Level 2 so each end extends outside the building model.

32.   Switch to the North elevation.

Notice that Level 2 looks the same as it did before you extended it using 2D extents in the South elevation.

33.   Save the file as *ex1-4rvt*.

Walls

Users will need to be familiar with the different parameters in walls. The user should also know which options are applied to walls and when those options are available.

Walls are system families. They are project-specific. This means the wall definition is only available in the active project. You can use Transfer Project Standards or Copy and Paste to copy a wall definition from one project to another.

On the Professional exam, you may be shown an image of a wall and asked to identify different wall properties.

Just as roofs, floors, and ceilings can consist of multiple horizontal layers, walls can consist of more than one vertical layer or region.

You can modify a wall type to define the structure of vertically compound walls using layers or regions.

Revit has several different wall types: Basic, Compound, Stacked, and Curtain

A Basic Wall is just what it sounds like, the standard "out of the box" wall style. This wall type may have several layers. For example, a brick exterior wall with a brick exterior layer, an air gap layer, a stud layer, an insulation layer, and a gypsum board layer.

A Compound wall is similar to a Basic wall. It also has layers, but one or more layers is divided into one or more regions, with each region being assigned a different material— for example, a wall that has an exterior layer that has concrete at the bottom and brick at the top.

A Stacked wall is two or more basic and/or compound walls that are stacked on top of each other. While Basic and Compound walls have a uniform thickness or width defined by the layers, a Stacked wall can have a variable thickness or width.

A Curtain wall is defined by a curtain grid. Mullions can be placed at the grid lines. Panels are placed in the spaces between the grid lines.

Exercise 1-5

Wall Options

Drawing Name: **i_firestation_basic_plan.rvt**
Estimated Time to Completion: 10 Minutes

Scope
Exploring the different wall options

Solution

1. Floor Plans
   Ground Floor
   Lower Roof
   Main Floor
   Main Roof
   Site
   T.O. Footing
   T.O. Parapet

   Activate the **Ground Floor** floor plan.

2. Zoom into the area where the green polygon is.

3. Select **Wall** from the Architecture tab on the ribbon.

4. Set the Wall Type to **Generic – 6″** in the Properties pane.

5. Set the Location Line to **Core Face:Exterior**.

6. Select the **Rectangle** tool on the Draw panel.

7. 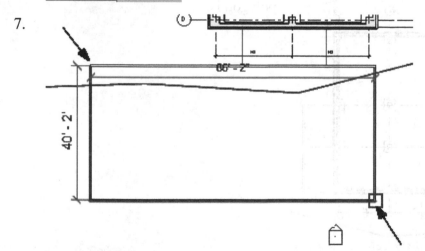 Select the two points indicated to place the rectangle.

8. Select **Wall** from the Architecture tab on the ribbon.

9. 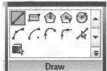 Select the **Line** tool from the Draw panel.

10. ──△────── Start the line at the midpoint of the lower horizontal wall.
Midpoint

11. Bring the line end up to the midpoint of the upper horizontal wall.
Left click to finish placing the wall.
Exit the Wall tool.

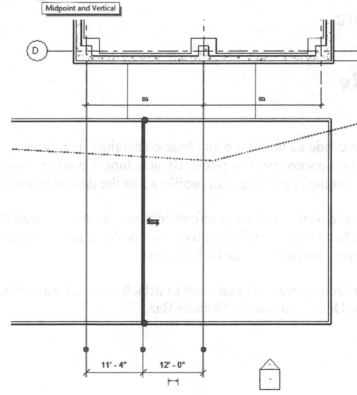

12. Select the vertical wall that you just placed.

Two temporary (listening) dimensions will appear.

Change the right dimension to **12′ [3600 mm]**.

13. 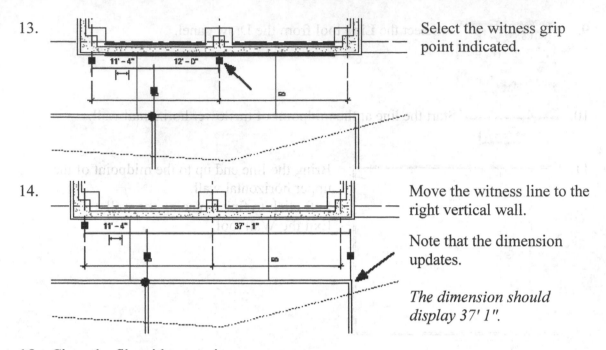 Select the witness grip point indicated.

14. Move the witness line to the right vertical wall.

Note that the dimension updates.

The dimension should display 37' 1".

15. Close the file without saving.

Attaching Walls

After placing a wall, you can override its initial top and base constraints by attaching its top or base to another element in the same vertical plane. By attaching a wall to another element, you avoid the need to manually edit the wall profile when the design changes.

The other element can be a floor, a roof, a ceiling, a reference plane, or another wall that is directly above or below. The height of the wall then increases or decreases as necessary to conform to the boundary represented by the attached element.

You can detach walls from elements as well. If you want to detach selected walls from all other elements at once, click Detach All on the Options Bar.

Exercise 1-6

Attaching Walls

Drawing Name: **i_Attach.rvt**
Estimated Time to Completion: 10 Minutes

Scope
Create a wall section view.
Attach a wall to a roof or floor.

Solution

1. Open *i_Attach.rvt*.

2.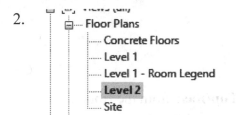
    Activate Level 2 Floor Plan.

3.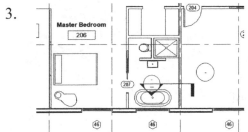
    Place a wall section as shown.

    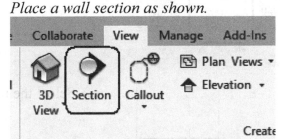

    Go to the **View** ribbon.
    Select the **Section** tool.

4.
    Set the view type to **Wall Section**.

5.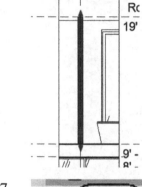

The first point selected will place the section head.
The second point selected will place the section tail.

Use the Flip controls if needed to orient the section head to face down/south.

Double left click on the section head to open the section view.

6.

Select the wall on Level 2.

7.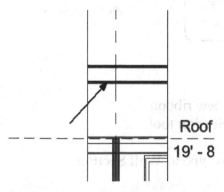

Select **Attach Top/Base** from the ribbon.

8.

Select the roof.

9.

What is the volume of the wall after it is attached to the roof?

Select the wall and then go to the Properties panel to determine the correct volume.
It should be 88.83 CF.

Exercise 1-7

Compound Walls

Drawing Name: **compound_walls.rvt**
Estimated Time to Completion: 60 Minutes

Scope
Defining a compound wall structure

A compound wall has multiple vertical layers and/or regions. A layer is assigned to each row with a constant thickness and extends the height of the wall. A region is any shape in the wall that is situated in one or more layers. The region may have a constant or variable thickness.

Solution

1. Open *compound_walls.rvt*.

2. Floor Plans
    Level 1
    Level 2
    Site   Activate **Level 1**.

3. Select the left vertical wall.

4. Properties
    Basic Wall
    Exterior - Brick on CMU
    Walls (1)   Edit Type   Select **Edit Type** on the Properties pane.

5. Duplicate...   Select **Duplicate**.

6. Name: Exterior - Concrete Foundation   Type **Exterior - Concrete Foundation**.
    Click **OK**.

7.

Type Parameters	
Parameter	Value
Construction	
Structure		Edit...	
Wrapping at Inserts	Do not wrap

Select **Edit Structure**.

8.

<< Preview

Expand the dialog by Clicking the **Preview** button.

9.

View: Floor Plan: Modify ty ▼   Preview >>
Floor Plan: Modify type attributes
Section: Modify type attributes

Switch the view to **Section: Modify Type attributes**.

10.

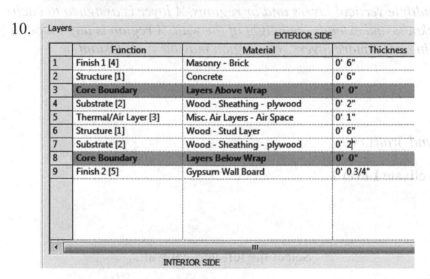

Layers
EXTERIOR SIDE

	Function	Material	Thickness
1	Finish 1 [4]	Masonry - Brick	0' 6"
2	Structure [1]	Concrete	0' 6"
3	**Core Boundary**	**Layers Above Wrap**	**0' 0"**
4	Substrate [2]	Wood - Sheathing - plywood	0' 2"
5	Thermal/Air Layer [3]	Misc. Air Layers - Air Space	0' 1"
6	Structure [1]	Wood - Stud Layer	0' 6"
7	Substrate [2]	Wood - Sheathing - plywood	0' 2"
8	**Core Boundary**	**Layers Below Wrap**	**0' 0"**
9	Finish 2 [5]	Gypsum Wall Board	0' 0 3/4"

INTERIOR SIDE

Add Layers as follows:
Layer 1: Finish 1 [4] Masonry - Brick 6″
Layer 2: Structure [1] Concrete 6″
Layer 3: Core Boundary
Layer 4: Substrate [2] Wood – Sheathing - plywood 2″
Layer 5: Thermal Air/Layer – Misc. Air Layers - Air Space 1″
Layer 6: Structure [1] Wood – Stud Layer 6″
Layer 7: Substrate [2] Wood – Sheathing - plywood 2″
Layer 8: Core Boundary
Layer 9: Finish 2 [5] Gypsum Wall Board 3/4″

11.

Split Region

Select **Split Region**.

12. 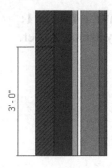 Cut the Layer 1: brick layer 3'-0" from the base.

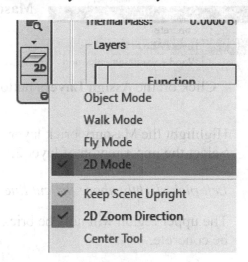

Toggle the 2D button to see the hatch patterns on the layers.

13.

	Function	Material
1	Finish 1 [4]	Masonry - Brick
2	Structure [1]	Concrete

Highlight the **Layer 2: Concrete** Layer.

14.  Pick on the **Assign Layers** button.

15. 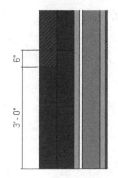 Select the lower region of the brick layer that was just split.

Left pick just below the cut line.

The upper region will now be brick and the lower region will be concrete.

It may take some practice before you are able to do this.

16. Split Region   Select **Split Region**.

17. Cut the Layer 2: concrete 3'-6" from the base.

18.

	Function	Material
1	Finish 1 [4]	Masonry - Brick
2	Structure [1]	Concrete
3	Core Boundary	Layers Above Wrap
4	Substrate [2]	Wood - Sheathing - plywood

EXTERIOR

Highlight the **Layer 1: Masonry Brick** Layer.

19.

Assign Layers

Click on the **Assign Layers** button.

20.

6"

3' - 0"

Highlight the Masonry brick layer.
Select the upper region of layer 2.

Left pick slightly above the cut line.

The upper region will now be brick and the lower region will be concrete.

It may take some practice before you are able to do this.

21.

Modify

Select **Modify**.

You can also use the Modify mode to correct the dimensions of the splits if needed.

22.

6"

3' - 0"

Select the base of the concrete Layer 1 component.

23.

A small lock will appear.
Click on the lock to unlock it.

24.

3' - 0"

Select the base of Layer 2: Concrete.
Click on the lock to unlock it.

25. Click **OK** to close the dialogs.

26.

Base is Attached	☐
Base Extension Distance	-3' 0"
Top Constraint	Unconnected
Unconnected Height	20' 0"
Top Offset	0' 0"

Select the wall with the Exterior - Concrete Foundation wall type.
In the Properties pane:
Set the Base Extension Distance to **-3' 0"**.
Left Click in the display window to release the selection.

27.

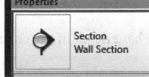

Set the display to Medium or Fine to see the wall layers.

28.

Activate the **View** ribbon.
Select the **Section** Tool from the Create panel.

29.

Set the section type to be a **Wall Section**.

30.

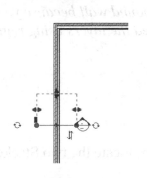

Place a small section on the wall you just defined.
Activate the section view.

31.

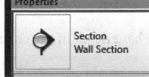

Set the display to Medium or Fine to see the wall layers.

32.

Note the concrete section is below the base level.
Select the wall.

33.

You can use the grips to adjust the base depth of the concrete section.

This type of wall is called a compound wall because you have split wall layers and modified the layers using regions.

34.

Floor Plans
Level 1
Level 2
Site

Activate **Level 1**.

35.

Stacked Wall
Exterior - Brick Over CMU w Metal Stud
Stacked Wall 1

In the Project Browser, locate the two Stacked Wall types.

These are the stacked walls available in the current project.

36.

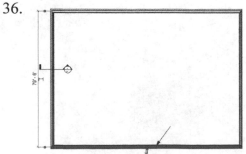

Select the south wall.

37.

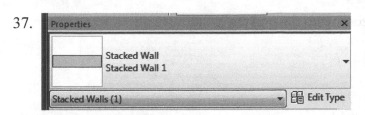

Switch the wall to **Stacked Wall 1** using the Type Selector on the Properties panel.

Select **Edit Type**.

38.

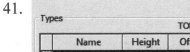

Select **Duplicate**.

39.

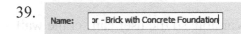

Rename **Exterior - Brick with Concrete Foundation**.
Click **OK**.

40.

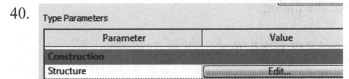

Select **Edit** Structure.

41.

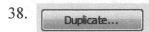

Note the stacked wall uses different layers going from Top to Base instead of Exterior to Interior.

Each layer is a wall type instead of a component material. These wall types are called subwalls.

42.

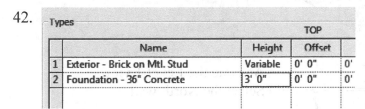

Change Layer 1 to **Exterior Brick on Mtl. Stud**.
Change Layer 2 to **Foundation - 36" Concrete**.
Set the Height of Layer 2 to **3'- 0"**.

43.

Select **Insert**. Position the new layer between the existing layers.

44.

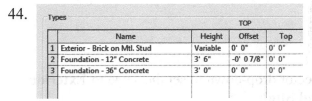

Set the new layer to:
Foundation- 12" Concrete.
Set the Height to **3' 6"**.
Set the Offset to **-7/8"**.

45. You can zoom into the preview window to check the offset value.

46. The height of the Top Layer is set to Variable so the user can set the wall height.
Click **OK** twice to exit the dialog.

Subwalls can be moved up or down the height of a stacked wall.

47. 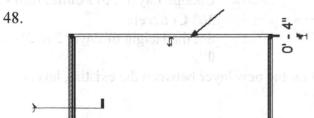 Switch to a 3D view so you can inspect the two wall types.
Note that when you hover the mouse over the first wall you defined, it displays as a Basic Wall.
The other wall displays as a Stacked Wall.

48. Activate the Level 1 view.
Select the North wall.

49. 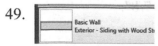 Use the Type Selector drop-down to set the Type to **Exterior - Siding with Wood Stud**.

50. 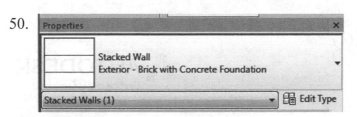 Select the South Wall (the stacked wall).
Select **Edit Type**.

51. 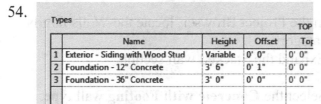 Select **Duplicate**.

52. Rename **Exterior - Siding with Concrete Foundation**.
Click **OK**.

53. Select **Edit** Structure.

54. 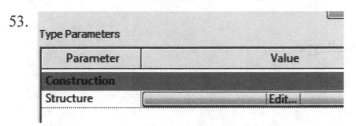 Set Layer 1 to the new wall type:
Exterior - Siding with Wood Stud.
Adjust the Offset for Layer 2:
Foundation -12" Concrete to **1"**.

55. You can zoom into the preview window to check the offset.
Click OK twice to close the dialogs.

56. 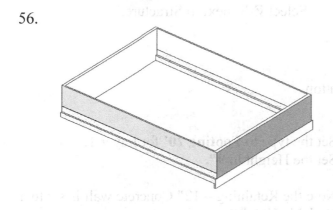 Switch to a 3D view.

How many stacked walls are there in the model?

How many basic walls?

What is the difference between a compound wall and a stacked wall?

Is a compound wall a basic wall or a stacked wall?

57. Close without saving.

Exercise 1-8

Stacked Walls

Drawing Name: **i_Footing.rvt**
Estimated Time to Completion: 10 Minutes

Scope
Defining a stacked wall.

A stacked wall uses more than one wall type.

Solution

1. 📂   Open *i_Footing.rvt*.

2. In the Project Browser, locate the Walls family category.
    Expand the Stacked wall section.

    Select the **Concrete with Footing** wall type.

3. Right click and select **Type Properties**.

4. Select **Edit** next to Structure.

5.    Select the **Insert** button.

6. Set the type to **Footing 20'** for Layer 1.
    Set the Height to **9".**

    Note the Retaining – 12" Concrete wall is set to a Variable Height.

7. 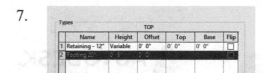 Highlight Layer 1.

Use the Down button to move the Footing 20'
below the Retaining – 12" Concrete.

8.

Select the Preview button to expand the dialog and see what the wall
looks like.

Click OK twice to close the dialog.

9. 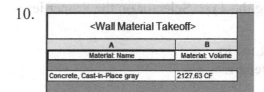 Locate the **Wall Material Takeoff** schedule in
the Project Browser.

Right click and select Open or double left click to
activate.

10.

<Wall Material Takeoff>	
A	B
Material: Name	Material: Volume
Concrete, Cast-in-Place gray	2127.63 CF

What is the material volume in cubic feet of the
Concrete, Cast-in-Place, gray material?

Did you get 2127.63 CF?

Exercise 1-9
Placing a Wall Sweep

Drawing Name: **walls.rvt**
Estimated Time to Completion: 20 Minutes

Scope
Placing a wall sweep

Solution

1. Activate **Level 1** Floor Plan.

   Floor Plans
   — Level 1
   — Level 2
   — Site

2. Select the **Wall** tool from the Architecture tab on the ribbon.

3. Set the wall type to **Exterior - Brick on Mtl. Stud** using the Type Selector on the Properties pane.

4. Set the Location Line to **Finish Face: Exterior**.

   Location Line: Finish Face: Ext
   Wall Centerline
   Core Centerline
   Finish Face: Exterior
   Finish Face: Interior
   Core Face: Exterior
   Core Face: Interior

5. Select the **Pick Line** mode from the Draw panel.
   Select the four green lines.

   Note that when you pick the lines, the side of the line you use determines which side of the line is used for the exterior side of the wall.

6.

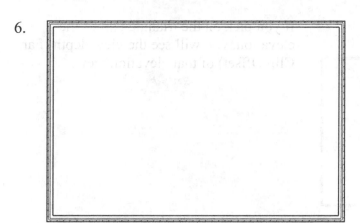

The lines should be aligned to the exterior side of the walls.

Set the Detail Level to **Medium**.

7.

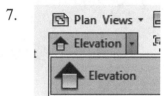

Activate the **View** ribbon.
Select the **Elevation** tool on the Create panel.

8.

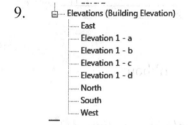

Place an elevation in the center of the room.
Right click and select **Cancel** to exit the command.

Click in the center of the square to select the elevation marker.

Place a check mark on each box to create an elevation for each interior wall.

9.

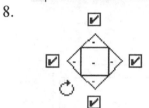

In the Project Browser, you will see that four elevation views have been created.

10.

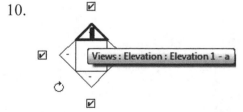

If you hover your mouse over a triangle, a tooltip will appear with the name of the linked view.

11.

Rename the elevation views to East Interior, North Interior, South Interior and West Interior.

*Clicking **F2** is a shortcut for Rename.*

12.

If you pick on the triangle part of the elevation, you will see the view depth (Far Clip Offset) of that elevation view.

13. Elevations (Building Elevation)
East
East Interior
North
North Interior
South
South Interior
West
West Interior

Activate the **South Interior** View.

14. Use the grips to extend the elevation view beyond the walls.

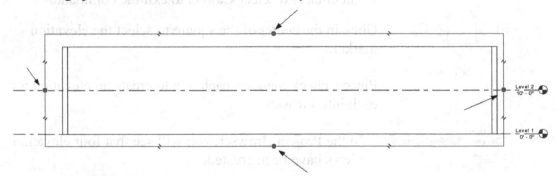

15. Architecture Structure

Wall Door Window C

Wall: Architectural

Wall: Structural

Wall by Face

Wall: Sweep

Activate the **Architecture** tab on the ribbon.
Select the **Wall Sweep** tool.

The Wall Sweep tool is only available in elevation, 3D or section views.

16. Left click to place the sweep so it is toward the top of the wall.
    Click ESC to exit the command.

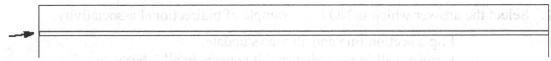

17.  Select the wall sweep.
    In the Properties pane, adjust the Offset from Level
    to **18′ 0″**.

18. Switch to a 3D view.

19. 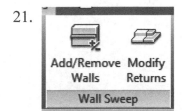 Select the top corners of the view cube to orient the view so you
    can see the wall sweep.

20. Select the wall sweep that was placed.
    It will highlight when selected.

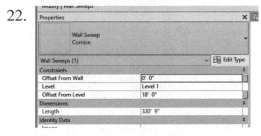

21. Select **Add/Remove Walls** from the ribbon.
    Select the other walls.
    Orbit around to inspect.

22. Select the wall sweep that was placed.
    Look in the Properties panel.

    What is the length of the wall sweep?

    You should see a value of 330' 9".

23. Save as *ex1-9.rvt*.

Certified User Practice Exam

1. Select the answer which is NOT an example of bidirectional associativity:

    A. Flip a section line and all views update.
    B. Draw a wall in plan view and it appears in all other views.
    C. Change an element type in a schedule and the change is displayed in the floor plan view as well.
    D. Flip a door orientation so the door swing is on the exterior of the building.

2. Select the answer which is NOT an example of a parametric relationship:

    A. A floor is attached to enclosing walls. When a wall moves, the floor updates so it remains connected to the walls.
    B. A series of windows are placed along a wall using an EQ dimension. The length of the wall is modified, and the windows remain equally spaced.
    C. A door is placed in a wall. The wall is moved, and the door remains constrained in the wall.
    D. A shared parameter file is loaded to the server.

3. Which tab does NOT appear on Revit's ribbon?

    A. Architecture
    B. Basics
    C. Insert
    D. View

4. Which item does NOT appear in the Project Browser?

    A. Families
    B. Groups
    C. Callouts
    D. Notes

5. Which is the most recently saved backup file?

    A. office.0001
    B. office.0002
    C. office.0003
    D. office.0004

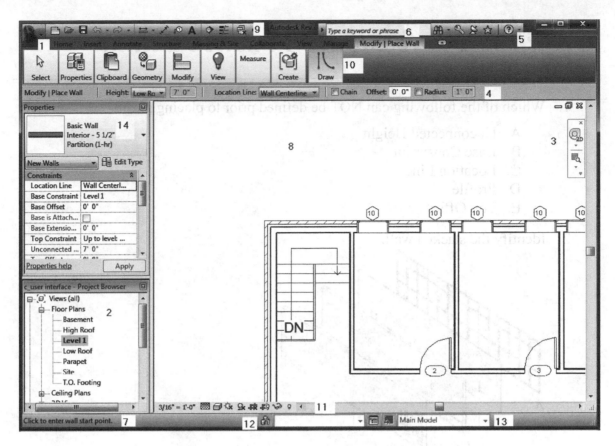

6. Match the numbers with their names.

View Control Bar	InfoCenter
Project Browser	Status Bar
Navigation Bar	Properties Pane
Options Bar	Application Menu
Design Options	Drawing Area
Help	Quick Access Toolbar
Ribbon	Worksets

Answers:

1) D; 2) D; 3) B; 4) D; 5) D; 6) 1- Application Menu, 2- Project Browser, 3- Navigation Bar, 4- Options Bar, 5- Help, 6-
InfoCenter, 7- Status Bar, 8- Drawing Area, 9- Quick Access Toolbar, 10- Ribbon, 11- View Control Bar,12- Worksets, 13-
Design Options

Certified Professional Practice Exam

1. Which of the following can NOT be defined prior to placing a wall?

    A. Unconnected Height
    B. Base Constraint
    C. Location Line
    D. Profile
    E. Top Offset

2. Identify the stacked wall.

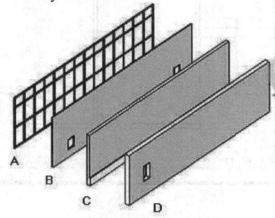

3. Walls are system families. Which name is NOT a wall family?

    A. BASIC
    B. STACKED
    C. CURTAIN
    D. COMPLICATED

4. Select the TWO that are wall type properties:

    A. COARSE FILL PATTERN
    B. LOCATION LINE
    C. TOP CONSTRAINT
    D. FUNCTION
    E. BASE CONSTRAINT

5. Select ONE item that is used when defining a compound wall:

    A. MATERIAL
    B. SWEEPS
    C. GRIDS
    D. LAYERS
    E. FILL PATTERN

6. Use this key to cycle through selections:

   A. TAB
   B. CTRL
   C. SHIFT
   D. ALT

7. The construction of a stacked wall is defined by different wall _____.

   A. Types
   B. Layers
   C. Regions
   D. Instances

8. To change the structure of a basic wall you must modify its:

   A. Type Parameters
   B. Instance Parameters
   C. Structural Usage
   D. Function

9. If a stacked wall is based on Level 1 but one of its subwalls is on Level 7, the base level for the subwall is Level _____.

   A. 7
   B. 1
   C. Unconnected
   D. Variable

Answers:
    1) D; 2) C; 3) D; 4) A & D; 5) D; 6) A; 7) A; 8) A; 9) B

Modifying Elements

This lesson addresses the following types of problems:

- Create and modify roofs
- Create and modify stairs
- Create and modify railings
- Create and modify floors
- Create and modify toposurfaces
- Create and modify columns

Building Elements are used to create a building design. There are five classes of building elements: host, component, datum, annotation, and view. Building elements fall into three categories: Model, View, and Annotation. To pass the User exam, users need to identify which category a building element falls in.

Each element falls into a category, such as wall, column, door, window, furniture, etc. Each category contains different families. Each family can have more than one type. The type is usually determined by the size or parameters assigned to that family.

These are very difficult concepts for many students, especially if they have been used to dealing with lines, circles, and arcs.

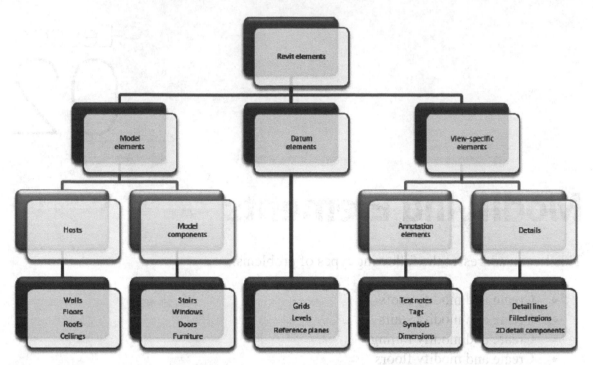

Revit elements are separated into three different types of elements: Model, Datum and View-specific. Users are expected to know if an element is model, datum or view specific.

Model elements are broken down into categories. A category might be a wall, window, door, or floor. If you look in the Project Browser, you will see a category called Families. If you expand the category, you will see the families for each category in the current project. Each family may contain multiple types.

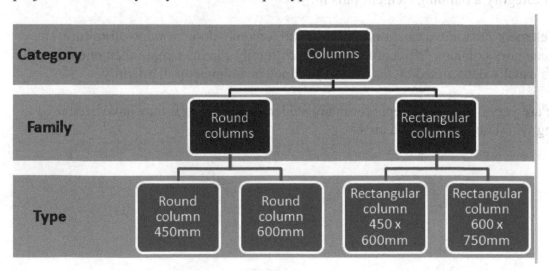

Every Revit file is considered a Project. A Revit project consists of the Project Environment, components, and views. The Project Environment is managed in the Project Browser.

Roofs

Roofs are system families. Roofs can be created from a building footprint, using an extrusion, and from a mass instance (converting a face to a roof).

Roof by footprint

- 2D closed-loop sketch of the roof perimeter

- Created when you select walls or draw lines in plan view

- Created at level of view in which it was sketched

- Height is controlled by Base Height Offset property

- Openings are defined by additional closed loops

- Slopes are defined when you apply a slope parameter to sketch lines

Roof by extrusion

- Open-loop sketch of the roof profile

- Created when you use lines and arcs to <u>sketch</u> the profile in an elevation view

- Height is controlled by the location of the sketch in elevation view

- Depth is calculated by Revit based on size of sketch, unless you specify <u>start and end points</u>.

Roofs are defined by material layers, similar to walls.

Exercise 2-1

AUTODESK.
Certified User

Creating a Roof by Footprint

Drawing Name: **i_roofs.rvt**
Estimated Time to Completion: 10 Minutes

Scope
Create a roof.
Determine the roof's volume.

Solution

1.

    Floor Plans
       Ground Floor
       Lower Roof
       Main Floor
       Main Roof
       Site
       T. O. Footing
       T. O. Parapet
    Ceiling Plans

Activate the **T.O. Parapet** floor plan.

2.

Select **Roof by Footprint** from the tab.

3.

Properties

Basic Roof
Wood Rafter 8" - Asphalt Shingle - Insulated

Basic Roof

Generic – 9"

Generic – 12"

Generic – 12" - Filled

Generic – 18"

Steel Truss - Insulation on Metal Deck - EPDM

Select **Steel Truss - Insulation on Metal Deck - EPDM** using the Type Selector on the Properties panel.

4.

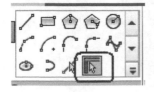

Select **Pick Walls** mode.

5. | ☐ Defines slope | Overhang: 0' 0" | ☐ Extend to wall core

On the Options bar,
uncheck **Defines slope**.
Set the Overhang to **0'-0"**.
Uncheck **Extend to Wall Core**.

6.   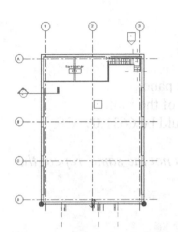   Select all the exterior walls.

7.   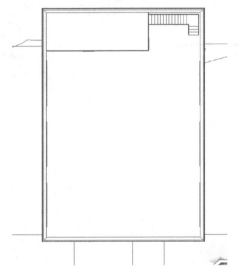   Use the **Trim** tool from the Modify panel to
create a closed boundary, if necessary.

8.   Align the boundary lines with the exterior face
of each wall.

9. 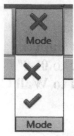 Select the **Green Check** on the Model panel to **Finish Roof**.

10.  Select the roof.

In the Properties panel,
note the volume of the roof.
The volume should be 6631.09. CF.

If your volume is not the same, repeat the exercise.

11. Close without saving.

Exercise 2-2

Creating a Roof by Extrusion

A AUTODESK.
Certified User

Drawing Name: **i_roofs_extrusion.rvt**
Estimated Time to Completion: 30 Minutes

Scope
Create a roof by extrusion.
Modify a roof.

Solution

1. 3D Views   Activate the **3D view.**
   {3D}

2.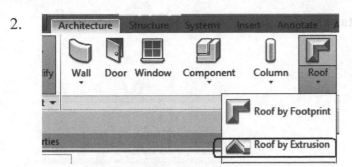

Activate the Architecture tab.
Select **Roof by Extrusion** under the Build panel.

3.

Enable **Name**.
Select **Roof Shape** from the list of reference planes.
Click **OK**.

4.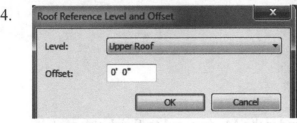

Select **Upper Roof** from the list.
Click **OK**.

5.

Select the **Show** tool from the Work Plane panel.
This will display the active work plane.

6.

Right click on the ViewCube's ring.
Select **Orient to a Plane**.

7.

Enable **Name**.
Select **Roof Shape** from the list of reference planes.
Click **OK**.

8. 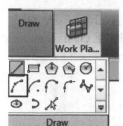 Select the **Start-End-Radius Arc** tool from the Draw panel.

9. 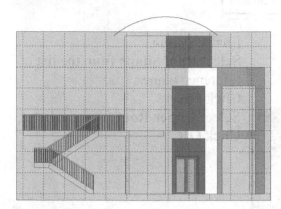 Draw an arc over the building as shown.

10. 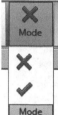 Select the **Green Check** under the Mode panel to finish the roof.

11. Switch to an isometric 3D view.

12. 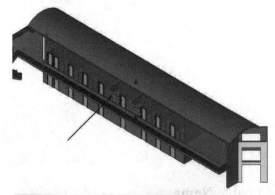 Hover over the wall below the roof.
Click TAB.
All connected walls will be selected.

13. 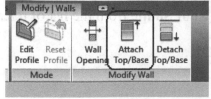 Select **Attach Top/Base** from the Modify Wall panel.
Then select the roof.
The walls will adjust to meet the roof.
Orbit the model to inspect the walls.

14. 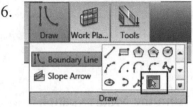 Activate the **Level 3** floor plan.

15. 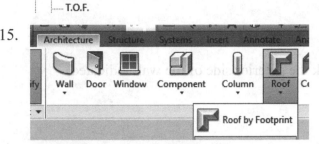 Activate the Architecture tab.
Select the **Roof By Footprint** tool from the Build panel.

16. 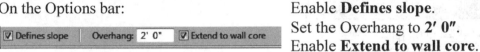 Select the **Pick Walls** tool from the Draw panel.

17. On the Options bar:

    | ☑ Defines slope | Overhang: 2' 0" | ☑ Extend to wall core |

    Enable **Defines slope**.
    Set the Overhang to **2' 0"**.
    Enable **Extend to wall core**.

18. Select the outside edge of the two walls indicated.

19. | ☐ Defines slope | Overhang: 2' 0" | ☑ Extend to wall core |

    Disable **Defines slope**.

20. Select the outside edge of the wall indicated.

21. | ☐ Defines slope | Overhang: 0' 0" | ☑ Extend to wall core |

    Set the Overhang to **0' 0"**.

22. 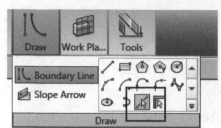   Select the **Pick Line** tool from the Draw panel.

23. 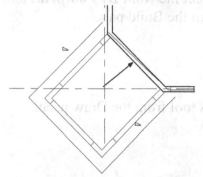   Pick the exterior side of the wall indicated.

24. 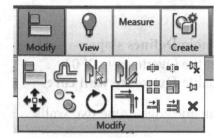   Select the **Trim** tool from the Modify panel.

25. 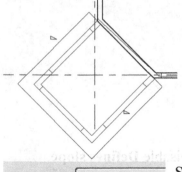   Trim the roof boundaries so they form a closed polygon.

26.    Set the Type to **Generic- 9″**. Click **OK**.

27.    Set the Slope to **6″/12″.**

28. 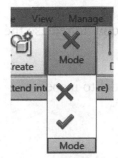 Verify that both the back and front boundary lines have no slope assigned.

If they have a slope symbol, select the back and front boundary and lines and uncheck **Defines Slope** on the Options bar.

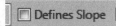

29.  Once the sketch is closed and trimmed properly, select the **Green Check** on the Mode panel.

30.  Return to a 3D view.

31. 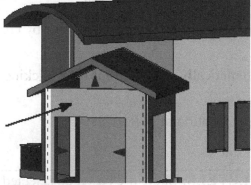 Hover over the front wall.
Click TAB to select all connected walls.
Left click to select the walls.

32. 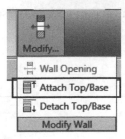 Select Attach Top/Base from the Modify Wall panel.

33.  Select the roof to attach the wall.

Roofs : Basic Roof : Generic - 9"

34. 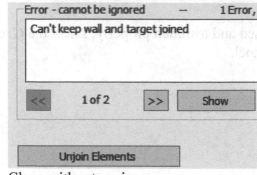 Click **Unjoin Elements**.

Error - cannot be ignored — 1 Error,

Can't keep wall and target joined

<< 1 of 2 >> Show

Unjoin Elements

Left click in the window to release the selection.

Orbit around the model to inspect the roofs.

35. Close without saving.

Stairs

Revit stairs are system families, similar to walls, ceilings, and floors. Railings are also system families, but the balusters and the railing profiles are loadable families.

A stair can consist of the following:

- Runs: straight, spiral, U-shaped, L-shaped, custom sketched run

- Landings: created automatically between runs or by picking 2 runs, or by creating a custom sketched landing

- Supports (side and center): created automatically with the runs or by picking a run or landing edge

- Railings: automatically generated during creation or placed later

Be familiar with all the properties that control stairs.

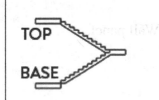

BASE AND TOP LEVELS: Stairs are based on selected levels that already exist in the project. You can add an offset on these levels if required.

	DESIRED STAIR HEIGHT: Total distance between the base and the top of the stairs, including offsets.
	DESIRED NUMBER OF RISERS: Automatically calculated by Revit, dividing Stair Height by Maximum Riser Height. You can change this number, which will modify the stair slope.
	ACTUAL NUMBER OF RISERS: The number of risers you modeled so far.
	MAXIMUM RISER HEIGHT: Riser height for your stair will never go above this value. This parameter is set on the stair type. Usually on par with code requirements.
	ACTUAL RISER HEIGHT: This distance is automatically calculated by Revit, dividing the Stair Height by the Desired Number of Risers.
	MINIMUM TREAD DEPTH: On the stair type, specify the minimum tread depth. When you start modeling your stair, you can go above this number, but not below.
	ACTUAL TREAD DEPTH: By default, this value is equal to minimum tread depth set in the stair type. However, you can set a bigger value if you want more depth.

	MINIMUM RUN WIDTH: Set on the stair type, you can specify the minimum run width. This does not include support (stringers).
	ACTUAL RUN WIDTH: By default, this will be the same as the minimum run width. You can set a higher value than the minimum, but a lower value will result in a Warning.

Exercise 2-3

Creating Stairs by Sketch

Drawing Name: **i_stairs.rvt**
Estimated Time to Completion: 20 Minutes

Scope
Place stairs using reference work planes.

Solution

1. 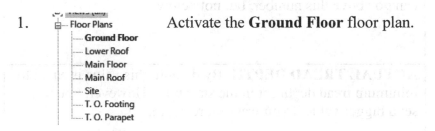   Activate the **Ground Floor** floor plan.

2. 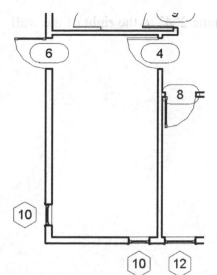 Zoom into the lower left corner of the building.

3. Select the **Reference Plane** tool from the Work Plane panel on the Architecture tab.

    ν **Ref Plane**

4. Offset: 2' 4″   Set the Offset to **2′ 4″** on the Options bar.

5. Draw   Select the **Pick Lines** tool from the Draw panel.

6. 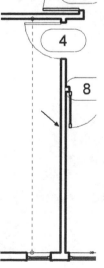 Place a vertical reference plane 2′ 4″ to the left of the wall where Door 4 is placed.

7.

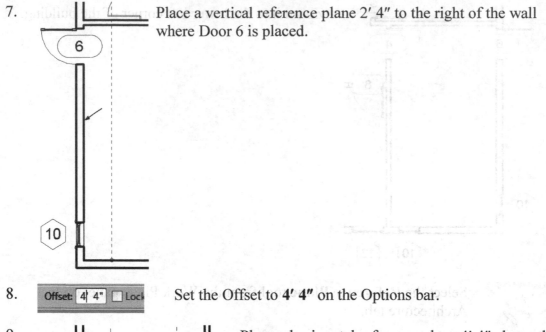

Place a vertical reference plane 2′ 4″ to the right of the wall where Door 6 is placed.

8.   Offset: 4′ 4″ ☐ Lock   Set the Offset to **4′ 4″** on the Options bar.

9.

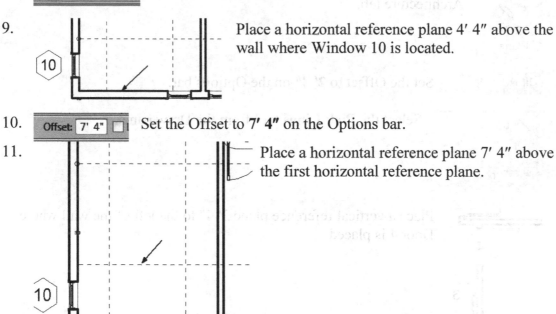

Place a horizontal reference plane 4′ 4″ above the wall where Window 10 is located.

10.   Offset: 7′ 4″ ☐   Set the Offset to **7′ 4″** on the Options bar.

11.

Place a horizontal reference plane 7′ 4″ above the first horizontal reference plane.

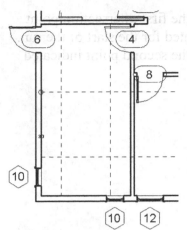

Four reference planes – two horizontal and two vertical should be placed in the room.

12.   Activate the Architecture tab.
Select the **Stair** tool from the Circulation panel.

13.   Select **Edit Type** from the Properties pane.

14.   Change the Maximum Riser Height to **8″**.
Set the stair width to **4′ 0″**.
Click **OK**.

Parameter	
Calculation Rules	
Maximum Riser Height	0' 8"
Minimum Tread Depth	0' 11"
Minimum Run Width	4' 0"
Calculation Rules	

15.   Set the desired number of risers to **16**.

Dimensions	
Desired Number of Risers	16
Actual Number of Risers	1
Actual Riser Height	0' 7 1/2"
Actual Tread Depth	0' 11"
Tread/Riser Start Number	1

16.   Select the **Run** tool from the Draw panel on the tab.

Run		Landing	Support
Components			

17.   Set the Location line to **Run: Center** on the Options bar.

Location Line: Run: Center

18.

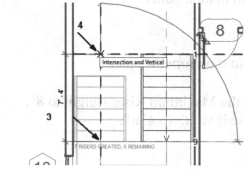

Pick the first intersection point indicated for the start of the run. Pick the second point indicated.

19.

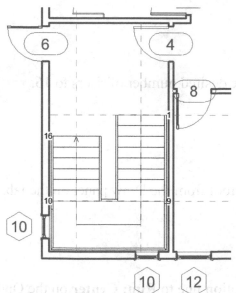

Moving clockwise, select the lower intersection point #3 and then the upper intersection point #4.

U-shaped stairs are placed.

20.

Select the **Green Check** on the Mode panel to **Finish Stairs**.

21.

Dimensions	
Width	4' 0"
Desired Number of Risers	16
Actual Number of Risers	16
Actual Riser Height	0' 7 1/2"
Actual Tread Depth	0' 11"

Select the stairs.

On the Properties pane,
locate the value for the Actual Tread Depth.
Locate the value for the Actual Riser Height.

22. Close without saving.

Exercise 2-4
Creating Stairs by Component

AUTODESK.
Certified User

Drawing Name: **i_stairs_component.rvt**
Estimated Time to Completion: 10 Minutes

Scope
Place stairs using the component method.

Solution

1. 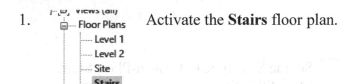 Activate the **Stairs** floor plan.

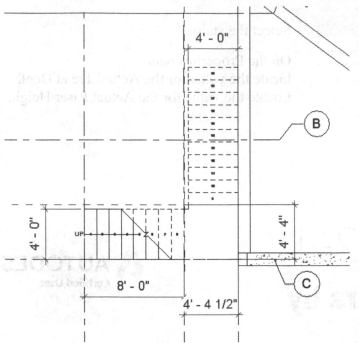

Stairs will be placed using the dimensions shown.

There are eight horizontal risers and 11 vertical risers.

Use the reference planes provided as guides.

Use the Cast in Place Monolithic Stair.

Set the Base Level to Level 1.

Set the Top Level to Level 2.

Set the Tread Depth to 1'-0".

Set the Desired Number of Risers to 19.

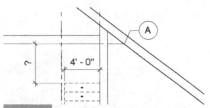

You need to determine the distance from the top riser labeled Riser 19 and Grid A.

2. Activate the Architecture tab.

   Select the **Stair** tool.

3. Set the Stair type to Cast-In-Place Stair: Monolithic Stair.

   Set the Base Level to Level 1.

   Set the Top Level to Level 2.

Cast-In-Place Stair Monolithic Stair	
Stair	
Constraints	
Base Level	Level 1
Base Offset	0' 0"
Top Level	Level 2
Top Offset	0' 0"
Desired Stair Height	11' 0"

4. Scroll down.

   Set the Tread Depth to 1'-0".

   Set the Desired Number of Risers to 19.

Dimensions	
Desired Number of Risers	19
Actual Number of Risers	1
Actual Riser Height	0' 6 243/256"
Actual Tread Depth	1' 0"
Tread/Riser Start Number	1
Identity Data	

5. 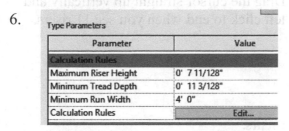   Select **Edit Type** on the Properties panel.

6. 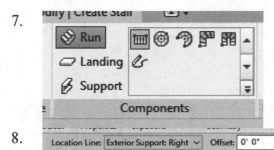   In the Type Parameters dialog:
Set the Minimum Run Width to **4' - 0"**.

Click **OK.**

7. 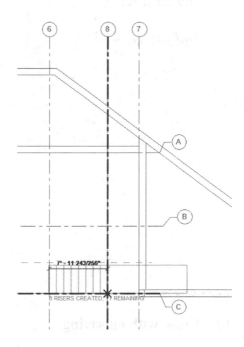   Enable **Run**.

8. Location Line: Exterior Support: Right ∨   Offset: 0' 0"   Actual Run Width: 4' 0"   ☑ Automatic Landing

On the Options bar: Set the Location Line: **Exterior Support Right**
Set the Offset to **0' 0"**. Set the Actual Run Width to **4' 0"**.

9. Start the stairs at C6 and end the run at C8.

This will place 8 risers.

10. 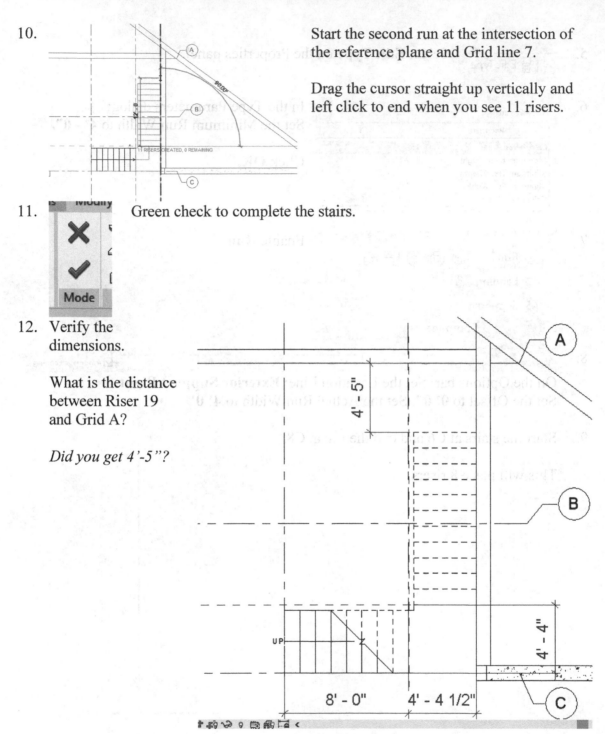 Start the second run at the intersection of the reference plane and Grid line 7.

Drag the cursor straight up vertically and left click to end when you see 11 risers.

11. Green check to complete the stairs.

12. Verify the dimensions.

What is the distance between Riser 19 and Grid A?

Did you get 4'-5"?

13. Close without saving.

Landings

Landings are placed automatically when you draw more than one run.

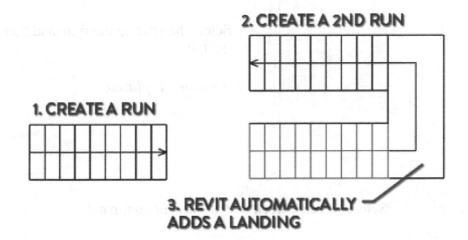

2. CREATE A 2ND RUN

1. CREATE A RUN

3. REVIT AUTOMATICALLY ADDS A LANDING

Exercise 2-5

Stair Landings

AUTODESK.
Certified User

Drawing Name: **landings.rvt**
Estimated Time to Completion: 10 Minutes

Scope
Create a stair landing

Solution

1. Floor Plans
    Ground Floor
    Ground Floor - Main Stairs
    Lower Roof
    Activate the **Ground Floor – Main Stairs** floor plan.

2. Stair
    Select **Stair** from the Architecture tab on the ribbon.

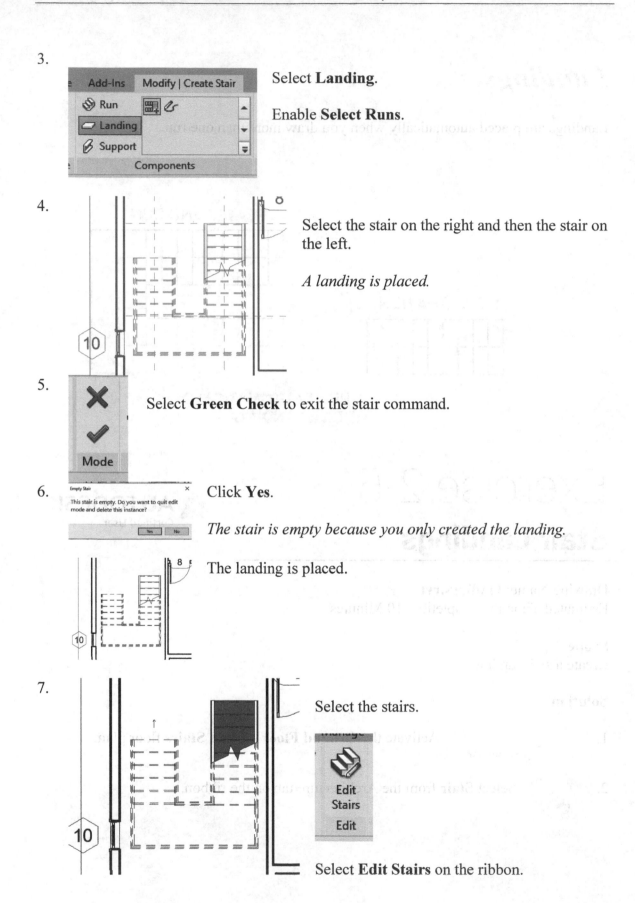

3. Select **Landing**.

Enable **Select Runs**.

4. Select the stair on the right and then the stair on the left.

A landing is placed.

5. Select **Green Check** to exit the stair command.

6. Click **Yes**.

The stair is empty because you only created the landing.

The landing is placed.

7. Select the stairs.

Select **Edit Stairs** on the ribbon.

8.

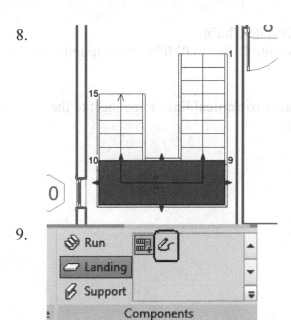

Select the landing that was just placed.

Right click and select Delete or click the DELETE key on the keyboard.

9.

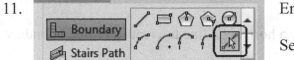

Select **Landing**.

Select **Edit Sketch**.

10.

Set the Offset to **0' 9"** on the Options bar.

11.

Enable **Boundary**.

Select the **Pick Line** tool.

12.

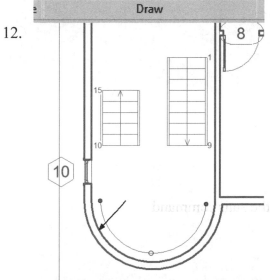

Select the inside of the arc on the inside face of the wall.

13.

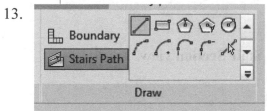

Select the **LINE** tool.

14.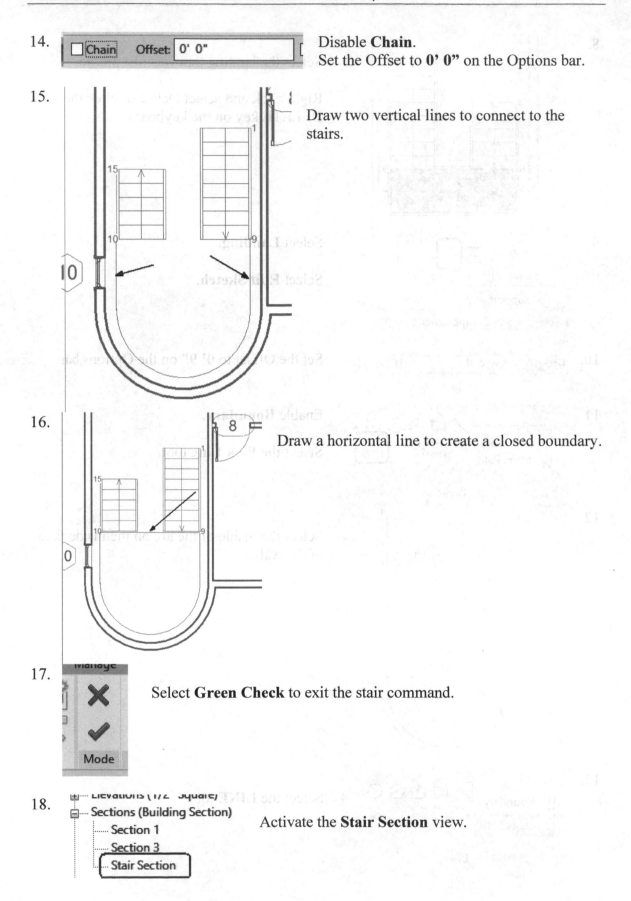

Disable **Chain**.
Set the Offset to **0' 0"** on the Options bar.

15.

Draw two vertical lines to connect to the stairs.

16.

Draw a horizontal line to create a closed boundary.

17.

Select **Green Check** to exit the stair command.

18.

Activate the **Stair Section** view.

19. You can see the landing.

Select the landing.

20. Adjust the Relative Height to **6' 0"**.

21. The landing position shifts.

22. Select **Green Check** to exit the stair command.

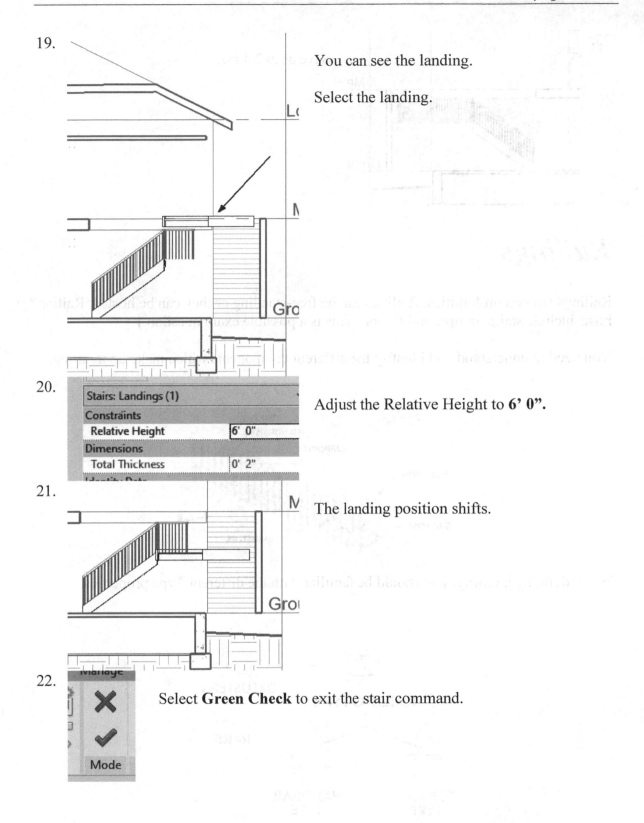

23.

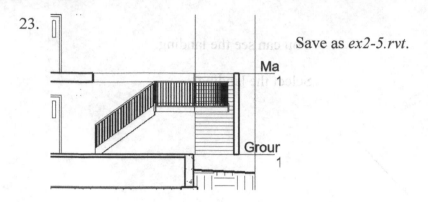

Save as *ex2-5.rvt*.

Railings

Railings are system families. Railings can be free-standing or they can be hosted. Railing hosts include stairs, ramps, and floors. (This is a possible exam question.)

You need to understand and identify the different components which make up a railing.

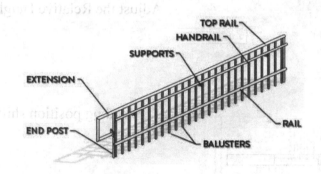

When defining a railing, you should be familiar with the different Type properties.

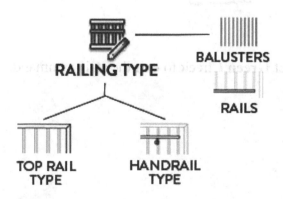

	TOP RAIL Top rail is the highest horizontal element of a railing. It is created by selecting a 2D profile and a height.
	HANDRAIL Handrail is an intermediate rail used for hands. They are linked to a wall or to a railing with Supports.
	INTERMEDIATE RAIL Any horizontal rail other than the Top Rail and the Handrail. Can be used to constrain balusters.
	RAIL 2D PROFILE Every Rail in Revit is an extrusion from a 2D Profile Family. Use default profiles for simple shapes, or create a custom one for fancy shapes.
	EXTENSION Use extension to add length to Top Rail or Handrail. The extension shape can be customized.
	SUPPORT The elements that connect the Handrail to the wall or to the railing.

Baluster Elements

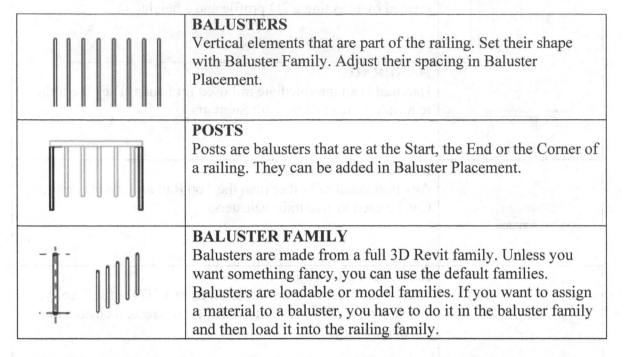	**BALUSTERS** Vertical elements that are part of the railing. Set their shape with Baluster Family. Adjust their spacing in Baluster Placement.
	POSTS Posts are balusters that are at the Start, the End or the Corner of a railing. They can be added in Baluster Placement.
	BALUSTER FAMILY Balusters are made from a full 3D Revit family. Unless you want something fancy, you can use the default families. Balusters are loadable or model families. If you want to assign a material to a baluster, you have to do it in the baluster family and then load it into the railing family.

You should be aware of where you click in the Railing's Type Properties to define different components.

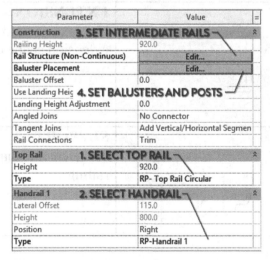

Exercise 2-6

Changing a Railing Profile

Drawing Name: **m_railing_family.rvt**
Estimated Time to Completion: 30 Minutes

Scope
Create a custom railing system family.
Create global parameters to manage materials.
Place railings on walls.

Solution

1. Go to the Insert tab.
   Select **Load Family**.

2. Browse to *Libraries\English\Profiles\Railings*.

3. Select the *M_Decorate Rail* family.
   Click **Open**.

4.   Site
     3D Views
          Approach
          From Yard
          Kitchen
          Living Room
          Parking Area
          Section Perspective
          Solar Analysis
          {3D}

Open the **Parking Area** 3D view.

5.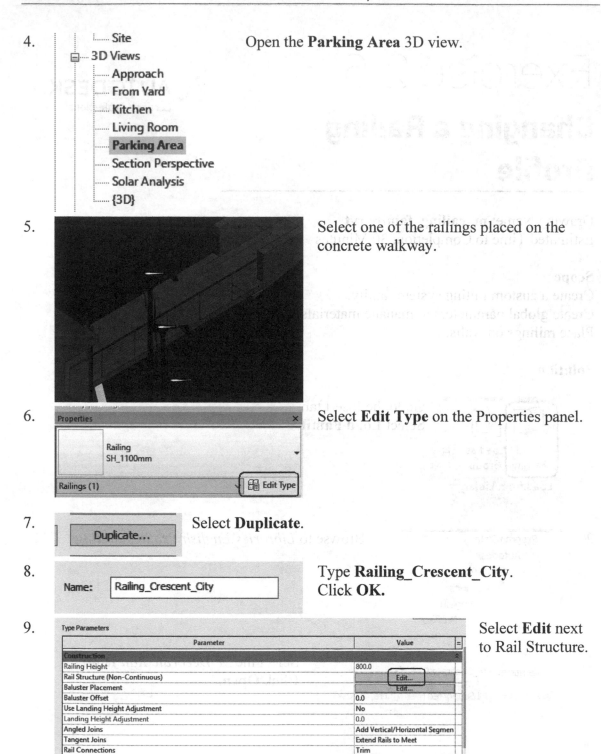

Select one of the railings placed on the concrete walkway.

6.   **Properties**

Railing
SH_1100mm

Railings (1)   **Edit Type**

Select **Edit Type** on the Properties panel.

7.   Duplicate...

Select **Duplicate**.

8.   Name: Railing_Crescent_City

Type **Railing_Crescent_City**.
Click **OK.**

9.   Type Parameters

Parameter	Value	=
Construction		
Railing Height	800.0	
Rail Structure (Non-Continuous)	Edit...	
Baluster Placement	Edit...	
Baluster Offset	0.0	
Use Landing Height Adjustment	No	
Landing Height Adjustment	0.0	
Angled Joins	Add Vertical/Horizontal Segmen	
Tangent Joins	Extend Rails to Meet	
Rail Connections	Trim	

Select **Edit** next to Rail Structure.

10. Delete all the rails except for **New Rail (1)**. Set the Height to **250.0**. Set the Profile to **M_Rectangular HandRail: 50 x 50 mm.** Set the Material to **Iron, Wrought.**
    You will have to import the material into the document.
    Click **OK**.

	Name	Height	Offset	Profile	
1	New Rail(1)	250.0	0.0	M_Rectangular Handrail : 50 x 50mm	Iron, Wrought

11.

Construction	
Railing Height	800.0
Rail Structure (Non-Continuous)	Edit...
Baluster Placement	Edit...
Baluster Offset	0.0
Use Landing Height Adjustment	No
Landing Height Adjustment	0.0
Angled Joins	Add Vertical/Horizontal Segments
Tangent Joins	Extend Rails to Meet
Rail Connections	Trim
Top Rail	

Select **Edit** next to Baluster Placement.

12. Set the Baluster Family to **Baluster –Square: 25 mm**. Set the Host to **Top Rail Element**. Set the Top Offset to **50.0**. Set Distance from Previous to **200.00.**

	Name	Baluster Family	Base	Base offset	Top	Top offset	Dist. from previous	Offset
1	Pattern start	N/A	N/A	N/A	N/A	N/A	N/A	N/A
2	Regular balust	M_Baluster - Square : 25mm	Host	0.0	Top Rail Eleme	50.0	200	0.0
3	Pattern end	N/A	N/A	N/A	N/A	N/A	0.0	N/A

13. Set the Start Post to **None**. Set the Corner Post to **None**. Set the End Post to **None**. Click **OK**.

	Name	Baluster Family	Base	Base offset	Top	Top offset	Space	Offset
1	Start Post	None	Host	-247.0	Top Rail Ele	0.0	0.0	0.0
2	Corner Post	None	Host	-247.0	Top Rail Ele	0.0	0.0	0.0
3	End Post	None	Host	-247.0	Top Rail Ele	0.0	0.0	0.0

14.

Rail Connections	Trim
Top Rail	
Use Top Rail	☑
Height	800.0
Type	Rectangular - 50x50mm

Set the Top Rail Height to **800.00**. Click **OK**.

15. 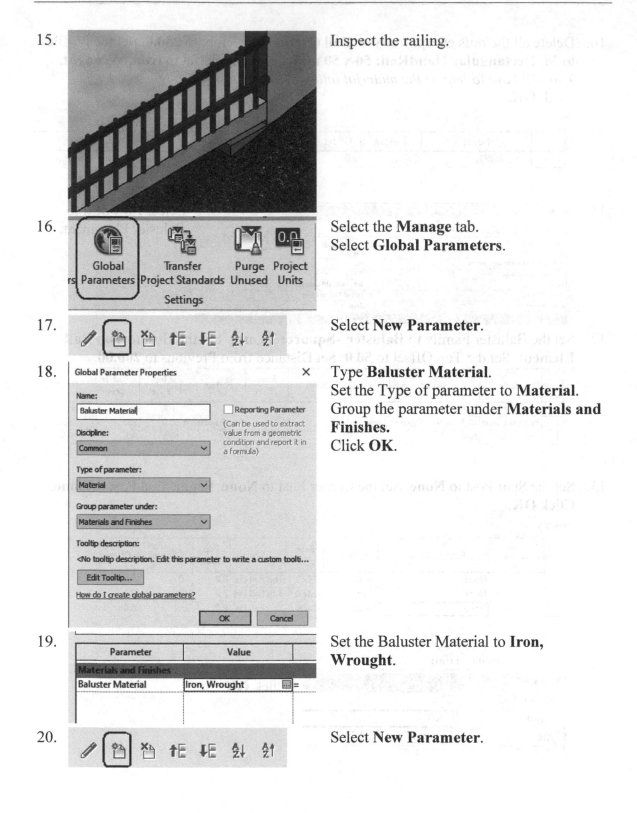   Inspect the railing.

16.   Select the **Manage** tab.
Select **Global Parameters**.

17.   Select **New Parameter**.

18.
Global Parameter Properties   ×

Name:

Baluster Material   ☐ Reporting Parameter

                                     (Can be used to extract
Discipline: value from a geometric
                                     condition and report it in
Common   ∨   a formula)

Type of parameter:

Material   ∨

Group parameter under:

Materials and Finishes   ∨

Tooltip description:

<No tooltip description. Edit this parameter to write a custom toolti...

Edit Tooltip...

How do I create global parameters?

                         OK   Cancel

Type **Baluster Material**.
Set the Type of parameter to **Material**.
Group the parameter under **Materials and Finishes.**
Click **OK**.

19.
Parameter	Value	
Materials and Finishes		
Baluster Material	Iron, Wrought	=

Set the Baluster Material to **Iron, Wrought**.

20.   Select **New Parameter**.

21.

Type **Railing Material**.
Set the Type of parameter to **Material**.
Group the parameter under **Materials and Finishes.**

Click **OK**.

22.

Set the Railing Material to **Iron, Wrought**.
Click **OK**.

23.

In the Browser,
Scroll down to the Families area.
Locate the Top Rail Type: Rectangular –
50 x 50 mm under the Railings category.
Right click and select **Type Propertie**s.

24. Assign the Material to **Iron, Wrought**.

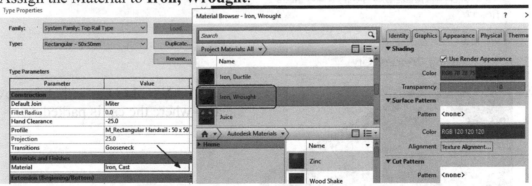

25.

Set the Profile for the Top Rail to:
M_Decorate Rail: 65 x 60 mm.

Click **OK**.

26.  The railing now appears to be a uniform material.

27. 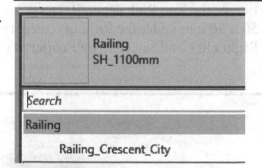 Select the second railing that was placed.

28. 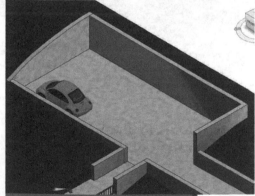 Use the Type Selector to change the railing to **Railing_Crescent_City.**

Railing
SH_1100mm

Search

Railing

Railing_Crescent_City

29. 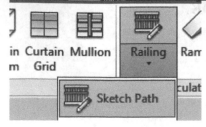 Use the PAN tool to move the view over to the area where the car is parked.

30. On the Architecture tab, select **Railing →Sketch Path.**

in Curtain Mullion   Railing   Ram
m   Grid

Sketch Path

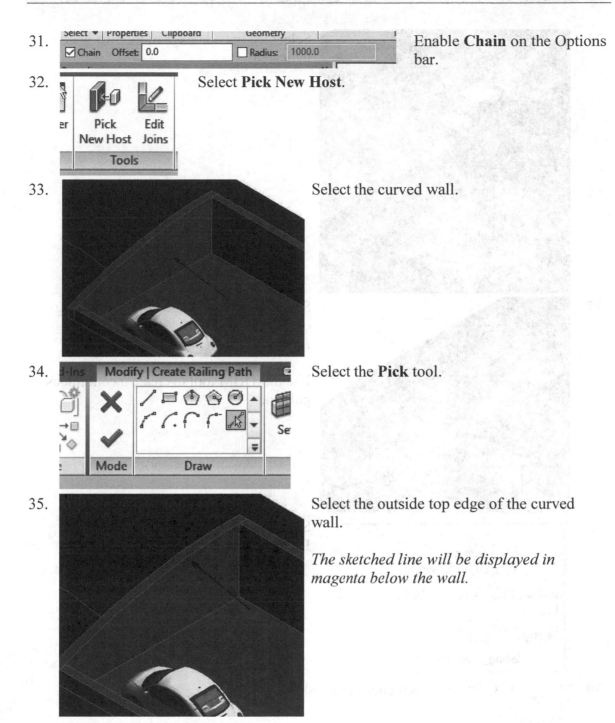

31. Enable **Chain** on the Options bar.

32. Select **Pick New Host**.

33. Select the curved wall.

34. Select the **Pick** tool.

35. Select the outside top edge of the curved wall.

    The sketched line will be displayed in magenta below the wall.

36.

Select the outside edge of the two walls indicated.

37.

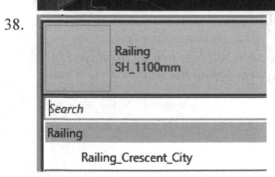

Select the outside edge of the remaining connecting walls.
Do not select the lower walls.

Use the TRIM tool to create a continuous polyline.

38.

Railing
SH_1100mm

Search

Railing

Railing_Crescent_City

Set the Railing Type to
Railing_Crescent_City.

39.

Mode

Click the green check to finish.

40.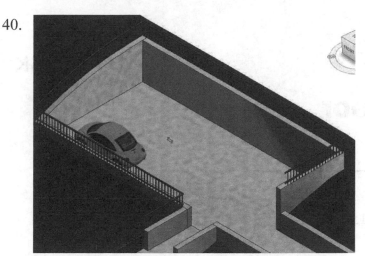

The railing is placed.

Select the railing.

What length is shown in the Properties pane?

Offset from Path	0.0
Dimensions	
Length	37829.1
Identity Data	

Save as *ex2-6.rvt*.

Floors

Floors are system families. They have layers, just like walls. You create floors by defining their boundaries, either by picking walls or using drawing tools. Floors are placed relative to levels. You do not need walls or a building pad to place a floor element.

Typically, you sketch a floor in a plan view, although you can use a 3D view if the work plane of the 3D view is set to the work plane of a plan view.

Floors are offset downward from the level on which they are sketched.

Exercise 2-7

Modifying a Floor Perimeter

Drawing Name: **i_floors.rvt**
Estimated Time to Completion: 15 Minutes

Scope
Modify a floor.
Determine the floor's perimeter.

Solution

1. ⊟ Floor Plans
     Ground Floor
     Lower Roof
     Main Floor
     Main Floor Admin Wing
     Main Roof
     Site
               Activate **the Main Floor Admin Wing** floor plan.

2. Window around the area indicated.

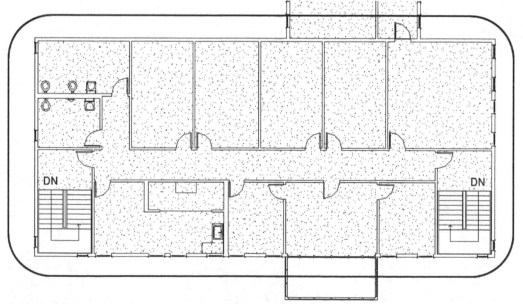

3. Select the **Filter** tool located in the lower right corner of the window.

4.    Select **Check None** to disable all the checks.

   Then check only the **Floors**.

   Click **OK**.

5. 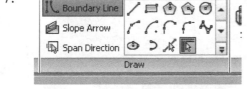   In the Properties pane,
   Note that the floor has a perimeter of 254' 8".

6. 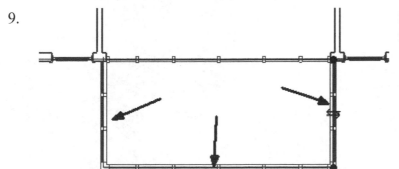   Select **Edit Boundary** under the Mode panel.

7.   Select **Pick Walls** mode under the Draw panel.

8.   Uncheck the Extend into wall (to core) option.

   Offset: 0' 0"   ☐ Extend into wall (to core)

9.   Select the three walls indicated.

10.   Select the **Trim** tool on the Modify panel from the tab.

11. 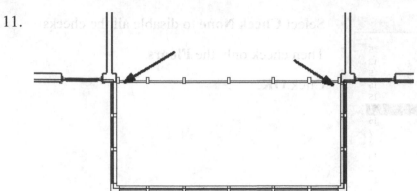 Trim the two corners indicated so that there is no wall in the section between the two vertical walls on the upper ends.

12.  Select the **Green Check** to **Finish Floor** from the tab.

13. Select **Don't attach**.

Attaching to floor   ×

Would you like walls that go up to this floor's level to attach to its bottom?

☐ Do not show me this message again   Attach   Don't attach

14.

Dimensions	
Slope	
Perimeter	270' 8"
Area	3390.71 SF
Volume	3602.63 CF
Thickness	1' 0 3/4"

Note that the floor has a perimeter of 270' 8". Click **OK** to close the dialog.

15. Close without saving.

Exercise 2-8
Modifying Floor Properties

Drawing Name: **floors_i.rvt**
Estimated Time to Completion: 5 Minutes

Scope
Modify a floor using Type Properties
Determine the floor's Elevation at Bottom

Solution

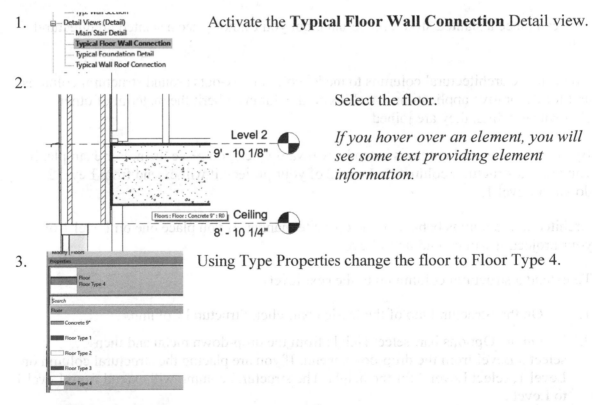

1. Activate the **Typical Floor Wall Connection** Detail view.

2. Select the floor.

    If you hover over an element, you will see some text providing element information.

3. Using Type Properties change the floor to Floor Type 4.

4. What is the Elevation at Bottom of the floor?

    It should read 8' 11 23/64"

Placing Columns

Although structural columns share many of the same properties as architectural columns, structural columns have additional properties defined by their configuration and industry standards which provide different behaviors.

Structural elements such as beams, braces, and isolated foundations join to structural columns; they do not join to architectural columns.

In addition, structural columns have an analytical model that is used for data exchange.

Typically, drawings or models received from an architect may contain a grid and architectural columns. You create structural columns by manually placing each column or by using the At Grids tool to add a column to selected grid intersections. In most cases, it is helpful to set up a grid before adding structural columns, as they snap to grid lines.

Structural columns can be created in plan or 3D views.

You can place a slanted structural column, but you cannot place a slanted architectural column.

You can use architectural columns to model column box-outs around structural columns and for decorative applications. Architectural columns inherit the material of other elements to which they are joined.

By default, structural columns extend downward when you place them. For example, if you place a structural column on Level 2 of your project, it will extend from Level 2 down to Level 1.

Architectural columns behave in the opposite manner. If you place one on Level 1 of your project, it will extend up to Level 2.

To extend a structural column up to the next level

1.   On the Structural tab of the Design bar, click Structural Column.

2.   On the Options bar, select Height from the drop-down menu and then select a Level from the drop-down menu. If you are placing the structural column on Level 1, select Level 2 for the height. The structural column will extend from Level 1 to Level 2.

Exercise 2-9

Placing Columns

Drawing Name: **columns.rvt**
Estimated Time to Completion: 30 Minutes

Scope
Place structural columns at grid lines.
Place an architectural column on top of a structural column.

Solution

1. [Views (all) / Floor Plans / **Level 1** / Level 2 / Site tree image]   Activate the **Level 1** floor plan.

2. [Column / Roof / Ceiling / Structural Column ribbon image]   Select **Structural Column** from the drop-down on the Architecture tab.

3. [W Shapes-Column / W10X33 / W10X49 image]   Select **W10X33** using the Type Selector.

4. [Modify / Height ∨ / Level 2 ∨ / 9' 0" Options bar image]   On the Options bar:
    Specify Height.
    Specify Level 2.

    Remember structural columns extend DOWN not up.

5. 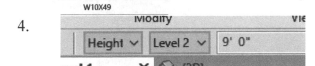   Left click on **At Grids**.

6.

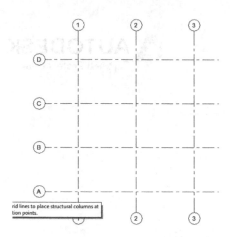

Hold down the CTL key.

Select Grids 1,2, 3, A,B,C, and D.

Select Finish on the tab.

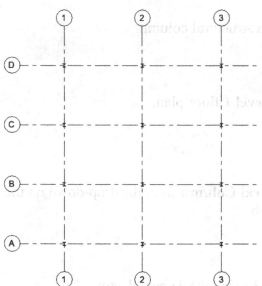

A structural column was placed at each grid intersection.

Click Cancel to exit the command.

7.

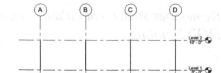

Activate the East elevation.

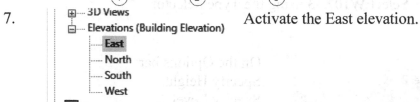

Notice that the structural columns were placed between Level 1 and Level 2.

8.

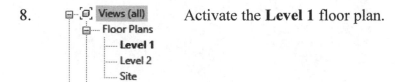

Activate the **Level 1** floor plan.

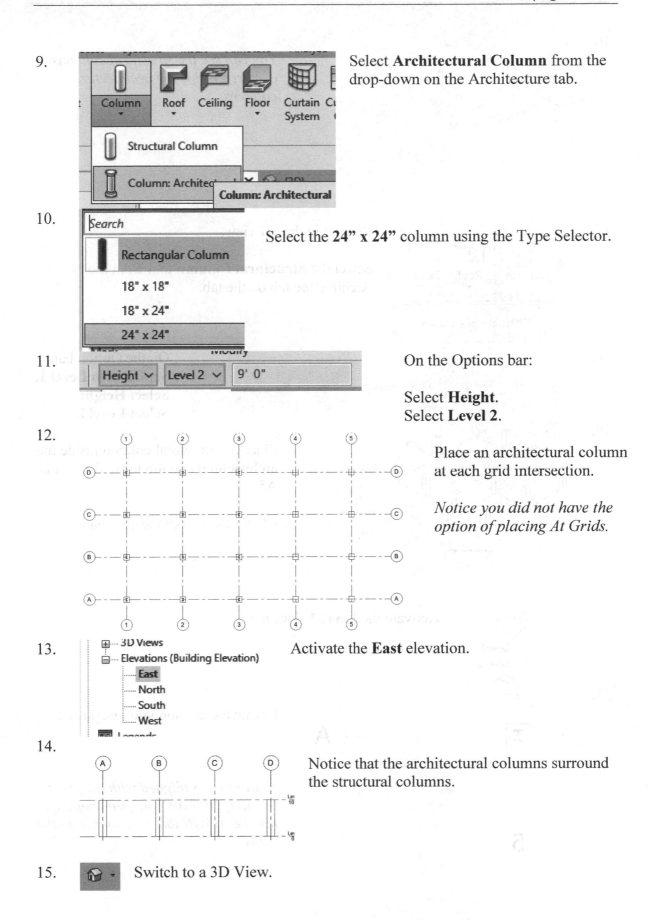

9.   Select **Architectural Column** from the drop-down on the Architecture tab.

10. Select the **24" x 24"** column using the Type Selector.

11. On the Options bar:

Select **Height**.
Select **Level 2**.

12. Place an architectural column at each grid intersection.

Notice you did not have the option of placing At Grids.

13. Activate the **East** elevation.

14. Notice that the architectural columns surround the structural columns.

15. Switch to a 3D View.

16.

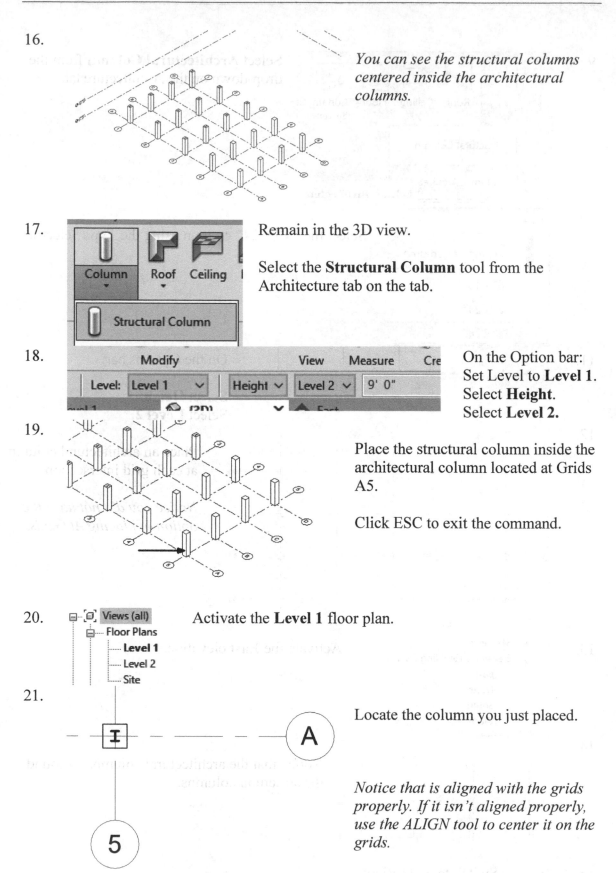

You can see the structural columns centered inside the architectural columns.

17.

Remain in the 3D view.

Select the **Structural Column** tool from the Architecture tab on the tab.

18.

On the Option bar:
Set Level to **Level 1**.
Select **Height**.
Select **Level 2.**

19.

Place the structural column inside the architectural column located at Grids A5.

Click ESC to exit the command.

20.

Activate the **Level 1** floor plan.

21.

Locate the column you just placed.

Notice that is aligned with the grids properly. If it isn't aligned properly, use the ALIGN tool to center it on the grids.

22. 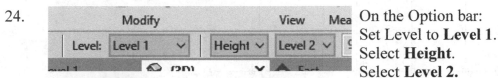 Switch to a 3D View.

23. Remain in the 3D view.

Select the **Structural Column** tool from the Architecture tab on the tab.

24. On the Option bar:
Set Level to **Level 1**.
Select **Height**.
Select **Level 2**.

25. 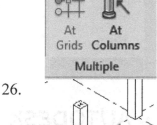 Enable **At Columns** on the ribbon.

26. 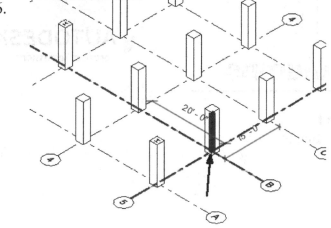 Hold down the CTL key.

Select the remaining columns on Grid 5.

27. 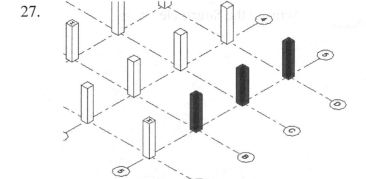  Select **Finish**.

Click **ESC** to cancel out of the command.

28. 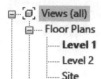 Activate the **Level 1** floor plan.

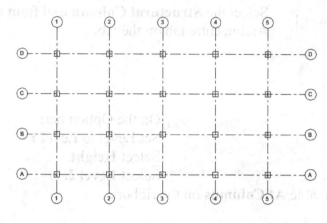

29. Save as *ex2-9.rvt*.

Exercise 2-10
Place Slanted Columns

Drawing Name: **slanted_column.rvt**
Estimated Time to Completion: 10 Minutes

Scope
Place a slanted structural column

Solution

1. Activate the **South** elevation.

   Elevations (Building Elevation)
   ├── East
   ├── North
   ├── **South**
   └── West

2.

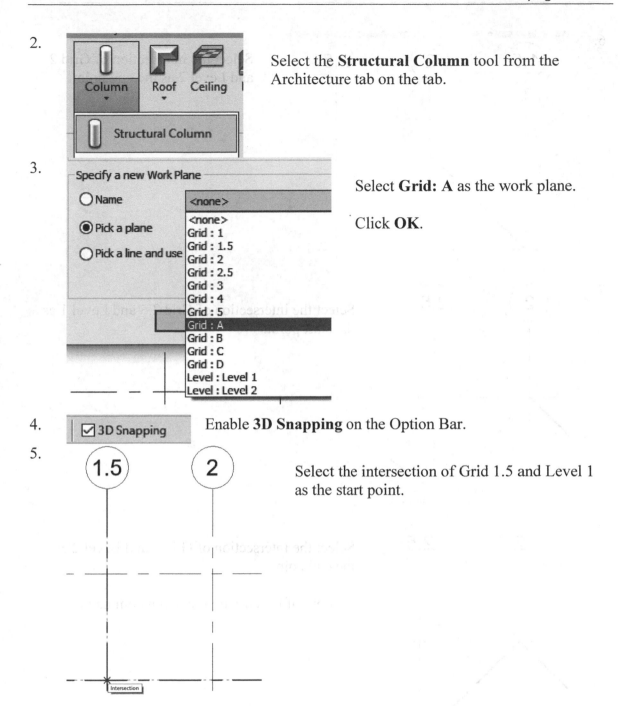

Select the **Structural Column** tool from the Architecture tab on the tab.

3.

Select **Grid: A** as the work plane.

Click **OK**.

4. ☑ 3D Snapping   Enable **3D Snapping** on the Option Bar.

5.

Select the intersection of Grid 1.5 and Level 1 as the start point.

6.

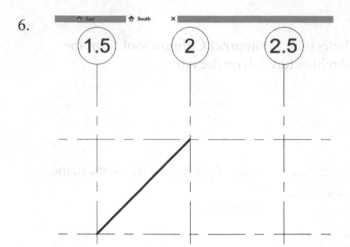

Select the intersection of Grid 2 and Level 2 as the end point.

7.

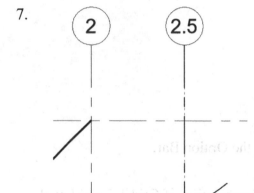

Select the intersection of Grid 2.5 and Level 1 as the start point.

8.

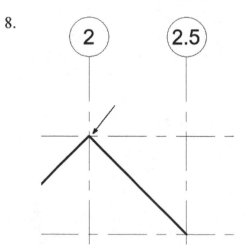

Select the intersection of Grid 2 and Level 2 as the end point.

Exit out of the structural column command.

9.   Switch to a 3D View.

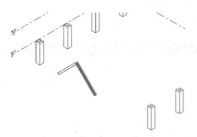

10.   Save as *ex2-10.rvt*.

Curtain Walls

A curtain wall is any exterior wall that is attached to the building structure and which does not carry the floor or roof loads of the building. Like walls, curtain walls are system families.

In common usage, curtain walls are often defined as thin, usually aluminum-framed walls containing in-fills of glass, metal panels, or thin stone. When you draw the curtain wall, a single panel is extended the length of the wall. If you create a curtain wall that has automatic curtain grids, the wall is subdivided into several panels.

In a curtain wall, grid lines define where the mullions are placed. Mullions are the structural elements that divide adjacent window units. You can modify a curtain wall by selecting the wall and right-clicking to access a context menu. The context menu provides several choices for manipulating the curtain wall, such as selecting panels and mullions.

Curtain Walls contains most properties of a basic wall. They have bottom and top constraints and their profile can be modified.

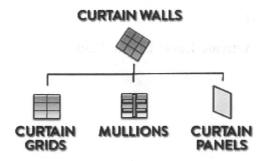

CURTAIN GRIDS

Curtain grids are divisions created on the walls. These divisions can be horizontal or vertical.

MULLIONS

Mullions are elements that can be created on each curtain grid segment, as well as on each curtain wall extremity.

CURTAIN PANELS

Curtain panels are rectangular elements located between each curtain grids.

Exercise 2-11:

Modify Elements Within a Curtain Wall

Drawing Name: **curtain_wall.rvt**
Estimated Time to Completion: 5 Minutes

This exercise reinforces the following skills:

- ❑ Curtain Wall
- ❑ Modify Wall Curtain Wall
- ❑ Elevations
- ❑ Grid Lines
- ❑ Load Family
- ❑ Properties
- ❑ Type Selector

1. Open *curtain_wall.rvt*.

2. 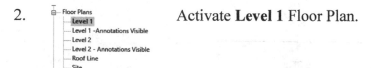 Activate **Level 1** Floor Plan.

3. Select the East Wall.

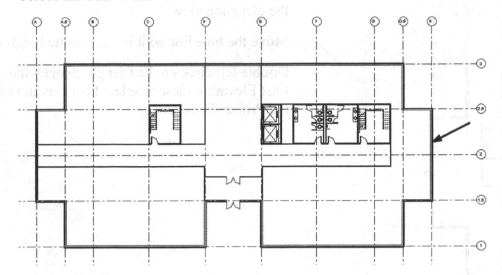

4.
    Select the **Storefront** Curtain Wall using the Type
    Selector on the Properties pane.

    Curtain Wall

    Curtain Wall 1

    Exterior Glazing

    Storefront

5.
Visibility
☑ Elevations
☑ Floor Tags
☑ Furniture System Tags

    Type **VV** to launch the Visibility/Graphics dialog.
    Open the **Annotations** tab.
    Enable **Elevations**.
    Click **OK**.

6.

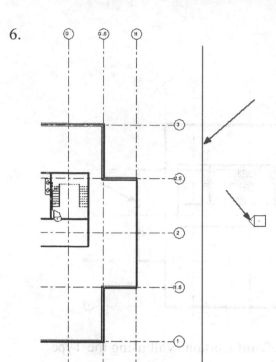

Left click on the triangle part of the elevation. A blue line will appear to indicate the depth of the elevation view.

Move the blue line so it is outside the building.

Double left click on the triangle to open the East Elevation view or select East elevation in the Project Browser.

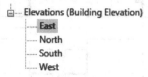

7.

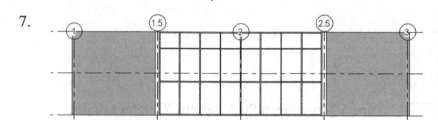

This is the east elevation view when the marker is outside the building and the depth is outside the building.

8.

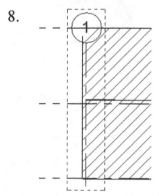

To adjust the grid lines, select the first grid line.

9.

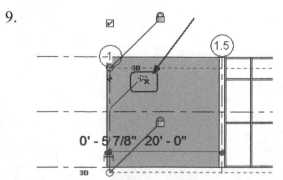

Unpin the grid line if it has a pin on it.

0' - 5 7/8"   20' - 0"

10.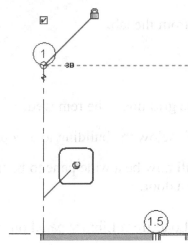

Select the grid line again.

Drag the bubble using the small circle above the building model.

11.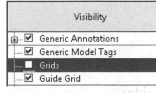

Left click to re-enable the pin.

12.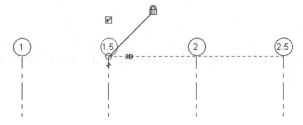

The group will adjust. Drag the remaining gridlines so the bubbles are aligned with the grid line group.

13.
Visibility
⊞ ☑ Generic Annotations
☑ Generic Model Tags
■ Grids
☑ Guide Grid

Type VV to launch the Visibility/Graphics dialog. Select the Annotations tab. Turn off visibility of grids.

Click **OK**.

14.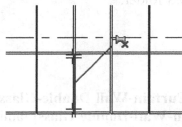

Select the center mullion. *Use the TAB key to cycle select.*

Unpin the selection.

Right click and select **Delete** or click the **Delete** key on your keyboard to delete.

15.  Select the grid line.

16. Select the **Add/Remove Segments** tool from the tab.

17.  Select the grid line to be removed.

Left click below the building to accept.

There will now be a wide pane to be used to create a door.

18. 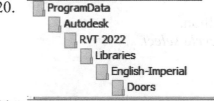 Select the **Load Family** tool from the Load from Library panel on the Insert tab.

19. Select the *Imperial Library* folder on the left pane of the Load Family dialog.

20. Browse to the *Doors* folder.

> ProgramData
> Autodesk
> RVT 2022
> Libraries
> English-Imperial
> Doors

21. Locate the **Door-Curtain-Wall-Double-Glass [M_ Door-Curtain-Wall-Double-Glass]** family from the *Doors* folder.

File name: Door-Curtain-Wall-Double-Glass

Files of type: All Supported Files (*.rfa, *.adsk)

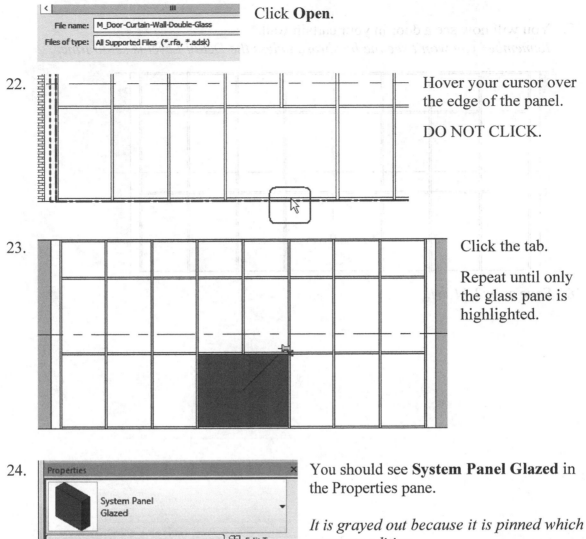

Click Open.

22. Hover your cursor over the edge of the panel.

DO NOT CLICK.

23. Click the tab.

Repeat until only the glass pane is highlighted.

24. You should see **System Panel Glazed** in the Properties pane.

It is grayed out because it is pinned which prevents editing.

25. 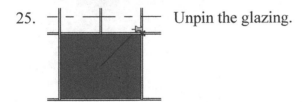 Unpin the glazing.

26. From the Properties pane:
Assign the **Door-Curtain-Wall-Double-Glass [M_ Door-Curtain-Wall-Double-Glass]** door to the selected element.

Left click to complete the command.

27. You will now see a door in your curtain wall.
    Remember you won't see the hardware unless the Detail Level is set to Fine.

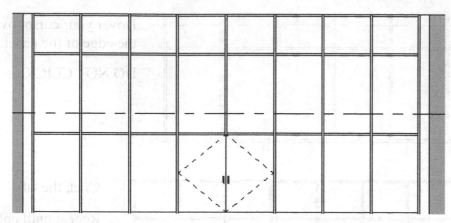

28. Save as *ex2-11.rvt*.

Certified User Practice Exam

1. The boundary of a floor is defined by:

    A. Model Lines
    B. Reference Lines
    C. Slab Edges
    D. Sketch Lines
    E. Drafting Lines

2. For the stairs shown, the vertical green lines represent:

    A. Risers
    B. Boundary Lines
    C. Stringers
    D. Run
    E. Railings

    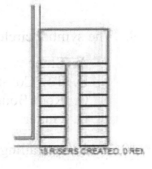

3. **True or False**: Stairs can be placed on multiple levels at a time.

4. **True or False**: You can change the direction of stairs using the flip arrows.

5. Identify the type of roof:

    A. Gable
    B. Hip
    C. Tar and Gravel
    D. Flat
    E. Shingled

    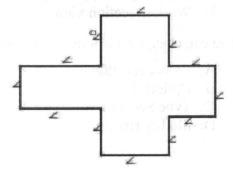

6. Name the three methods used to place a roof:

    A. By face
    B. By profile
    C. By footprint
    D. By extrusion
    E. By sketch

7. **True or False**: The extrusion of a roof can extend only in a positive direction from the selected work plane.

8. **True or False**: When attaching a wall to a roof, the profile of the wall does not change.

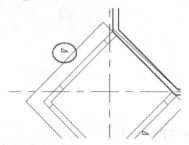

9. The symbol circled indicates:

    A. Slope
    B. Top Constraint
    C. Roof Boundary
    D. Gutter

10. To place a railing:

    A. Draw a path
    B. Select a host element
    C. Click to place
    D. Work in section view

11. You can change a Generic 12" Floor to a Wood Joist 10" Floor using the:

    A. Properties filter
    B. Options Bar
    C. Type Selector
    D. Modify ribbon

ANSWERS: 1) D; 2) B; 3) T; 4) T; 5) B; 6) 1,C,D; 7) T; 8) F; 9) A; 10) A; 11) C

Certified Professional Practice Exam

1. To create a stair landing, use the _____ tool.

    A. Run
    B. Boundary
    C. Riser
    D. Landing

2. To attach the top of a wall to a roof:

    A. Drag the top of the wall by grips
    B. Select the wall, check on Attach Top/Base and select roof
    C. Define wall attachments in instance properties
    D. Use the ALIGN tool

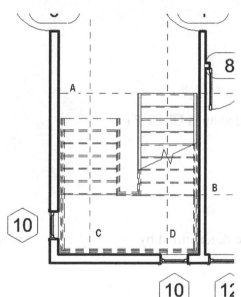

3.
    Identify the reference plane used to place the landing.

4.

Type Parameters				Top Rail	
Parameter		Value		Use Top Rail	☑
Construction				Height	1100.0
Railing Height	1100.0			Type C	Rectangular - 50x50mm
Rail Structure (Non-Continuous) A		Edit...		Handrail 1	
Baluster Placement B		Edit...		Lateral Offset	
Baluster Offset	0.0			Height	
Use Landing Height Adjustment	☐			Position	None
Landing Height Adjustment	0.0			Type D	<None>
Angled Joins	Add Vertical/Horizontal Segments			Handrail 2	
Tangent Joins	Extend Rails to Meet			Lateral Offset	
				Height	
				Position	None
				Type E	<None>
				Identity Data	

Select the area where you would click to change the profile used for the Top Rail.

5. Railings can be hosted by the following: (select all that apply)
   A. Slab edges
   B. Roofs
   C. Walls
   D. Stairs
   E. Ramps
   F. Floors

6. A Stair Run can be all the following shapes EXCEPT:

   A. Straight
   B. Spiral
   C. U-Shaped
   D. Closed

7. The shapes of rails and balusters are determined by _____, which must be loaded in the project.

   A. Profile families
   B. Type Properties
   C. System families
   D. In-place masses

8.   True or False

You can place an architectural column inside a structural column.

9.   True or False

You can place a slanted architectural column.

10. In a curtain wall, grid lines define where the _____ are placed.

      A. curtain panels
      B. curtain wall doors
      C. curtain walls
      D. mullions

ANSWERS: 1) D; 2) B; 3) B; 4) C; 5) C,D,E, & F; 6) D; 7) A; 8) F; 9) F; 10) D

Lesson

03

Modeling

This lesson addresses the following User and Professional exam questions:

- Managing Family Types
- Editing a model element's material
- Creating and modifying masses
- System and Component families
- Edit Room-Aware Families
- Creating and modifying toposurfaces
- Creating and modifying building pads
- Identifying family properties

Family Types

Using the Family Types tool, you can create many types (sizes) for a family.

To do this, you need to have labeled the dimensions and created the parameters that are going to vary.

Each family type has a set of properties (parameters) that includes the labeled dimensions and their values. You can also add values for standard parameters of the family (such as Material, Model, Manufacturer, Type Mark, and others).

To manage families and family types, use the shortcut menu in the Project Browser.

1. In the Project Browser, under Families, locate the desired family or type.

2. To manage families and family types, do the following:

If you want to...	then...
change type properties	right-click a type, and click Type Properties. As an alternative, double-click a type, or select it and Click Enter.
rename a family or type	right-click the family or type, and click Rename. As an alternative, select the type and Click F2.
add a type from an existing type	right-click a type, and click Duplicate. Enter a name for the type. The new type displays in the type list. Double-click the new type to open the Type Properties dialog, and change properties for the new type.
add a new type	right-click a family, and click New Type. Enter a name for the type. The new type displays in the list. Double-click the new type to open the Type Properties dialog, and define properties for the new type.
copy and paste a type into another project	right-click a type, and click Copy to Clipboard. Open the other project and, in the drawing area, Click Ctrl+V to paste it.
reload a family	right-click a family, and click Reload. Navigate to the location of the updated family, select it, and click Open.
edit a family in the Family Editor	right-click a family, and click Edit. The family opens in the Family Editor.
delete a family or type	right-click a family or type, and click Delete from Project. In addition to deleting the family or type from the project, this function deletes instances of matching types that exist in the model. Note: You cannot delete the last type in a system family.

Exercise 3-1

Create a New Family Type

Drawing Name: **family_type.rvt**
Estimated Time to Completion: 5 Minutes

Scope
Create a new family type.

Solution

1. 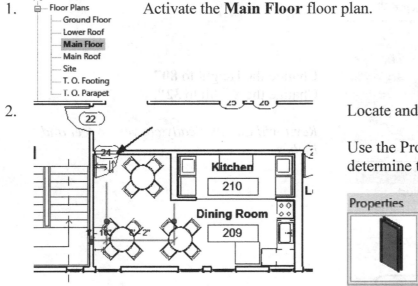 Activate the **Main Floor** floor plan.

2. Locate and select Door 24.

   Use the Properties panel to determine the door type.

3. In the Project Browser:

   Locate the Single-Flush door family.

   Expand to see the different door types available.

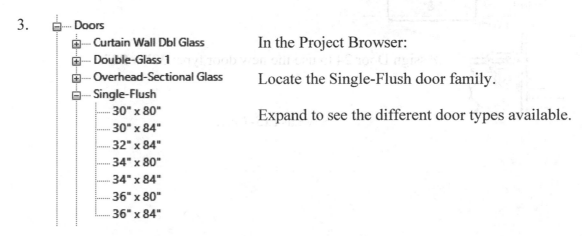

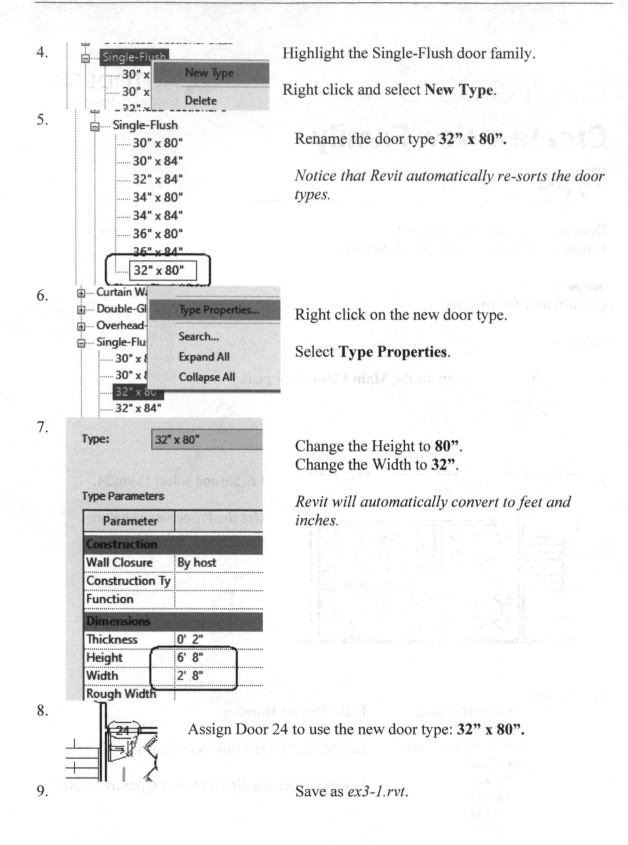

4.   Highlight the Single-Flush door family.

Right click and select **New Type**.

5.   Rename the door type **32" x 80"**.

Notice that Revit automatically re-sorts the door types.

6.   Right click on the new door type.

Select **Type Properties**.

7.   Change the Height to **80"**.
Change the Width to **32"**.

Revit will automatically convert to feet and inches.

8.   Assign Door 24 to use the new door type: **32" x 80"**.

9.   Save as *ex3-1.rvt*.

Exercise 3-2

Indentifying a Family

Drawing Name: **i_firestation_elem.rvt**
Estimated Time to Completion: 5 Minutes

Scope
Identify different elements and their families.

Solution

1. Open *i_firestation_elem.rvt.*

2. 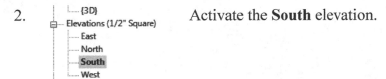 Activate the **South** elevation.

3. 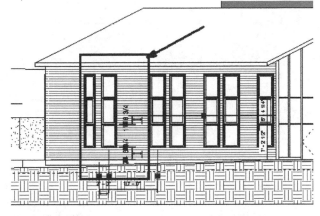 Select the second window from the left.

4. 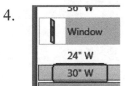 Select the **30″ W** from the window type list in the Properties pane.

5. 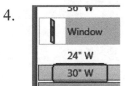 Use the Measure tool from the Quick Access toolbar to check the distance between the outside edges of the two left windows.

6.

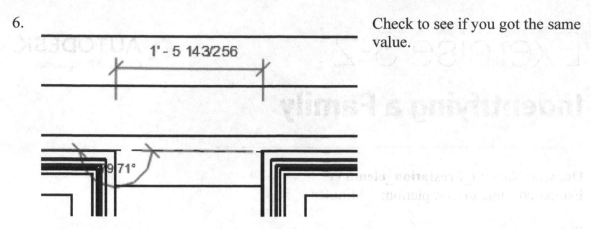

Check to see if you got the same value.

7. Close without saving.

Exercise 3-3

Changing the Material of a Door

Drawing Name: **door_material.rvt**
Estimated Time to Completion: 15 Minutes

Scope
Change the material of a door.

Solution

1. Open *door_material.rvt*.

2. 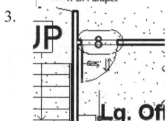 Activate the **Ground Floor** floor plan.

3. 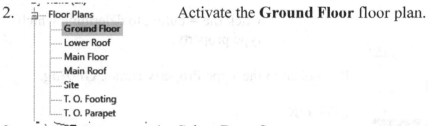 Select **Door 8**.

4. 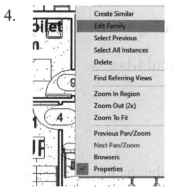 Right click and select **Edit Family**.

5.

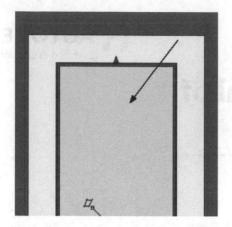

Select the glass pane.

6.

Visibility/Graphics Overrides	Edit...
Materials and Finishes	
Material	Glass
Identity Data	

On the Properties Panel, you see the Material is set to Glass, but it is grayed out.

It is grayed out because it is linked to a type property.

Click the = button to determine which type property.

7.

Search parameters

| <none> |
| Door Handle Material |
| Frame Material |
| Glass Stop Material |
| Glazing |
| Panel Material |

It is linked to the Type Property named **Glazing**.

Click **OK**.

8. Select **Family Types** on the tab.

...perties

9.

Dry Wall Frame (default)	☐
Materials and Finishes	
Panel Material	Wood - Birch
Frame Material	Metal - Paint Finish -
Glazing	Glass
Glass Stop Material	Metal - Paint Finish -
Door Handle Material	Aluminum

Notice that Glazing is set to **Glass.**

Click on the word **Glass**.

10.

Dry Wall Frame (default) ☐

Materials and Finishes	
Panel Material	Wood - Birch
Frame Material	Metal - Paint Finish -
Glazing	Glass
Glass Stop Material	Metal - Paint Finish -
Door Handle Material	Aluminum

Click on the **...** button that appears when you clicked on the word Glass.

11.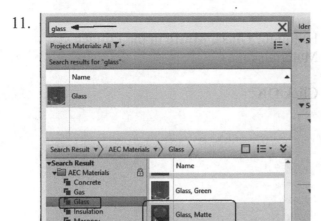

Type **glass** in the search field.

There is one Glass material available in the active project.

Locate the **Glass, Matte** material in the Materials library.

12.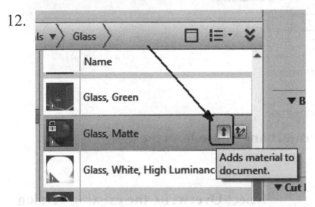

Use the Up arrow to add the **Glass, Matte** material to the active project.

13.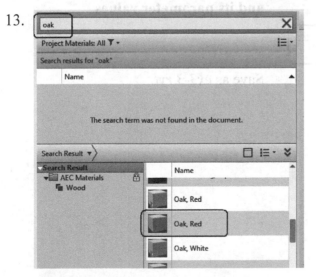

Type **oak** in the search field.

There are no oak materials available in the active project.

Locate the **Oak, Red** material in the Materials library.

14.

Use the Up arrow to add the **Oak, Red** material to the active project.

15. 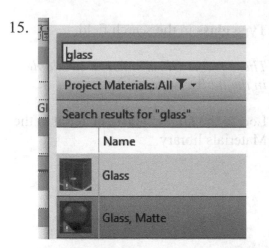 Highlight **Glass, Matte** in the Project Materials area.

Click **OK**.

16. 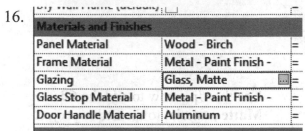 Verify that **Glass, Matte** is assigned to Glazing.

Materials and Finishes		
Panel Material	Wood - Birch	=
Frame Material	Metal - Paint Finish -	=
Glazing	Glass, Matte	=
Glass Stop Material	Metal - Paint Finish -	=
Door Handle Material	Aluminum	=

Click **OK**.

17. 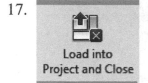 Save the file.

Select **Load into Project and Close** from the tab.

18. Select **Overwrite the existing version and its parameter values**.

→ Overwrite the existing version

→ Overwrite the existing version and its parameter values

19. Save as *ex3-3.rvt*.

Masses

You can create masses within a project (in-place masses) or outside of a project (loadable mass families).

Masses are used in a conceptual design environment to get an idea of the shape, footprint, and volume of a building.

Masses can be nested – that is, you can create a mass and then add or remove material from the mass to modify the shape or size.

In a project, you can join one mass to another mass to modify the floor area value. You can create schedules to determine the gross volume, floor area, and surface area of the building model.

Masses can be affected by phases, design options and worksets (used in collaboration).

Masses are considered an advanced topic and there is usually at least one massing question on the professional exam. In one question, you will be shown an image of a mass and asked how that mass was created. Another question might be on mass floors.

The Mass exercises in this text provide you with a good solid understanding of how to create and manipulate masses.

Masses start with one or more closed profile sketches and then you add height or thickness to the sketch, converting it to a mass. In order to modify the sketch, you have to open the mass up for editing. The sketch is nested below the mass. If you have a mass which consists of more than one mass, such as a base mass, then a void/hole, you have to open the mass for editing, then select the mass element to be edited and open that feature. This can be confusing for students as they have to go down the nested elements to get to the one they want to change.

Exercise 3-4

Create a Mass Using Revolve

Drawing Name: **mass_revolve.rvt**
Estimated Time to Completion: 20 Minutes

Scope
Create a mass using revolve.

Solution

1. Open *mass_revolve.rvt*.

2.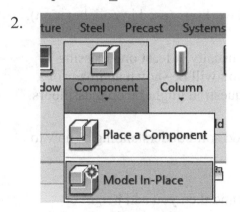

   On the Architecture tab:

   Select Model In-Place.

   Model In-Place allows you to create a mass object on the fly. This is a system family.

3.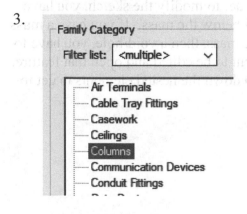

   Select **Columns**.

   Click **OK**.

4. Type **Column – Grecian**.

Click **OK**.

Name: Column - Grecian

OK

5. Select **Reference Plane** on the tab.

Reference Plane

6. Draw a horizontal and a vertical plane.

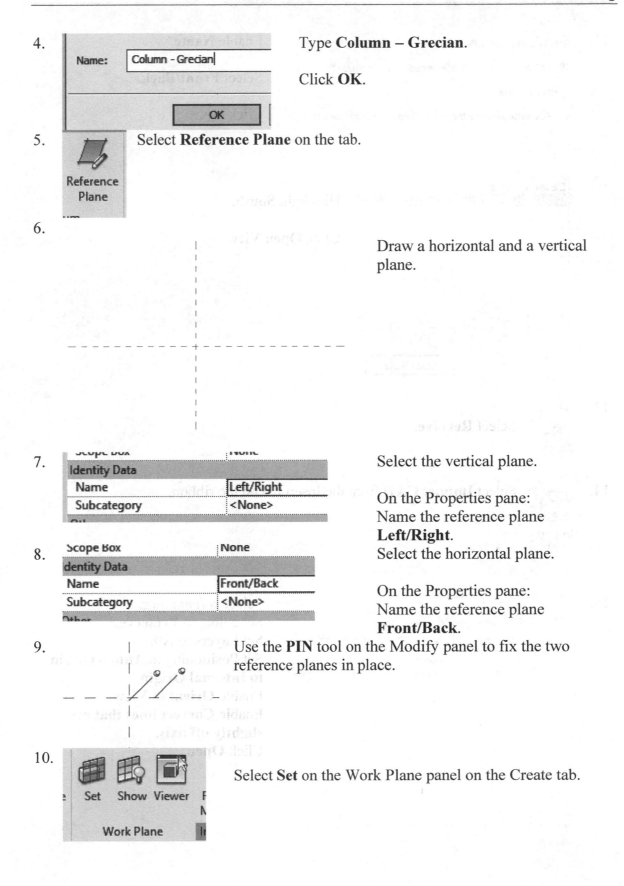

7. Select the vertical plane.

On the Properties pane:
Name the reference plane
Left/Right.

Identity Data	
Name	Left/Right
Subcategory	<None>

8. Select the horizontal plane.

On the Properties pane:
Name the reference plane
Front/Back.

Identity Data	
Name	Front/Back
Subcategory	<None>

9. Use the **PIN** tool on the Modify panel to fix the two reference planes in place.

10. Select **Set** on the Work Plane panel on the Create tab.

Set Show Viewer

Work Plane

11.

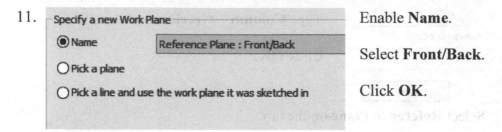

Enable **Name**.

Select **Front/Back**.

Click **OK**.

12.

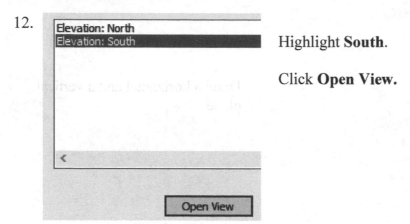

Highlight **South**.

Click **Open View.**

13.

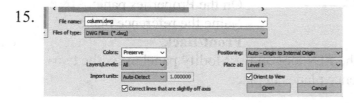

Select **Revolve**.

14.

Select **Import CAD** from the Insert tab on the ribbon.

15.

Select *column.dwg.*
Set colors to **Preserve.**
Set Layers to **All.**
Set Positioning to **Auto – Origin to Internal Origin**
Enable **Orient to View**
Enable **Correct lines that are slightly off axis**.
Click **Open**.

16.

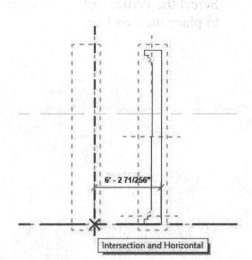

Use the **MOVE** tool to position the profile's right edge aligned with the vertical reference plane.

17.

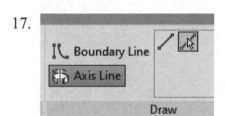

Highlight **Axis Line**.

Select the **Pick Line** tool.

18.

Select the vertical reference plane to place the axis line.

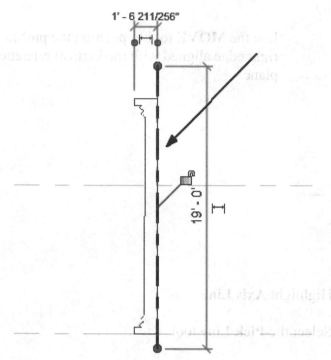

1' - 6 211/256"

19' - 0'

19.   Switch to a **3D** view.

20.   In the Properties pane:

Materials and Finishes	
Material	Concrete, Cast-in-Place gray

Set the Material to **Concrete, Cast-in-Place gray.**

21.

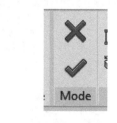

Click the **Green Check** to complete the revolve.

22.

The column is created.

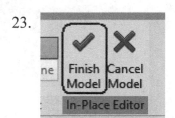

23.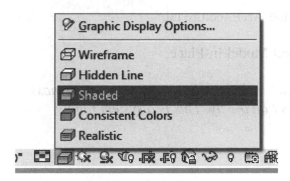

Select **Finish Model** on the tab.

24.

Change the display to **Shaded**.

✐ **G**raphic Display Options...

❐ Wireframe

❐ Hidden Line

❐ Shaded

❐ Consistent Colors

❐ Realistic

25.

Save as *ex3-4.rvt*.

Exercise 3-5

Create a Mass Using Blend

Drawing Name: **mass_blend.rvt**
Estimated Time to Completion: 15 Minutes

Scope
Create a mass using blend.

Solution

1. Open *mass_blend.rvt*.

2. 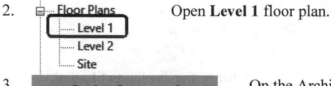 Open **Level 1** floor plan.

3. 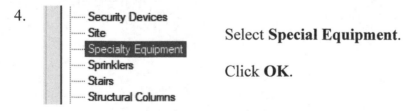 On the Architecture tab:

    Select Model In-Place.

    Model In-Place allows you to create a mass object on the fly. This is a system family.

4. 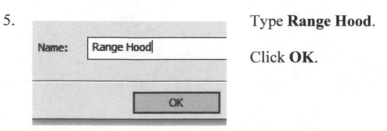 Select **Special Equipment**.

    Click **OK**.

5. Type **Range Hood**.

    Name: Range Hood

    Click **OK**.

6. Select the **Blend** tool on the tab.

7.  Select the **Modify|Create Blend Base Boundary** tab on the tab.

Select the **rectangle** tool.

8.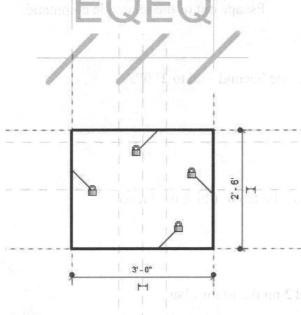

Draw the rectangle so it lies on top of the outside planes.

Enable the locks for each side of the rectangle.

This constrains the bottom profile to those planes.

Escape out of the rectangle command.

9. Select **Edit Top**.

10. 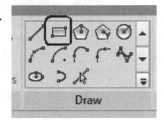 Select the **rectangle** tool.

11.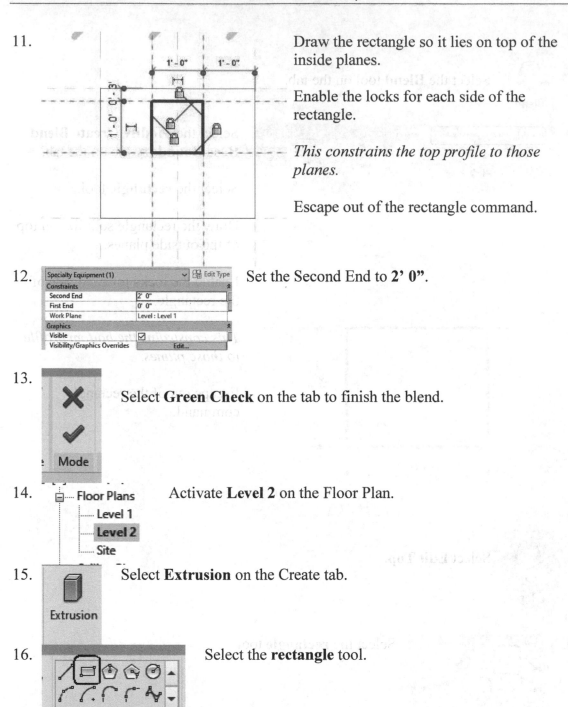

Draw the rectangle so it lies on top of the inside planes.

Enable the locks for each side of the rectangle.

This constrains the top profile to those planes.

Escape out of the rectangle command.

12. Set the Second End to **2' 0"**.

13. Select **Green Check** on the tab to finish the blend.

14. Activate **Level 2** on the Floor Plan.

15. Select **Extrusion** on the Create tab.

16. Select the **rectangle** tool.

17.

Draw the rectangle so it lies on top of the inside planes.

Enable the locks for each side of the rectangle.

This constrains the top profile to those planes.

18.

Constraints	
Extrusion End	3' 0"
Extrusion Start	0' 0"
Work Plane	Level : Level 2

On the Properties pane:

Set the Extrusion End to **3' 0"**.

19.

Select **Green Check** on the tab to finish the extrusion.

20.

Select **Green Check** on the tab to finish the model.

Finish Cancel
Model Model

In-Place Editor

21.

Switch to a 3D view so you can inspect your range hood.

Save as *ex3-5.rvt*.

Exercise 3-6

Create a Mass Using a Swept Blend

Drawing Name: **mass_swept_blend.rvt**
Estimated Time to Completion: 20 Minutes

Scope
Create a mass using a swept blend.

Solution

1. Open *mass_swept_blend.rvt*.

2.
   ┆········ [SD]
   ╞···· Elevations (Building Elevation)
   ┆······ **Downspout-West**
   ┆······ East
   ┆······ North
   ┆······ South
   ┆······ West
   —

   Open the **Downspout-West** elevation view.

3. 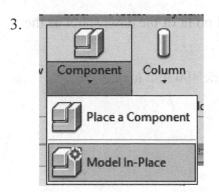 Select **Model In-Place** from the Architecture tab on the tab.

4. 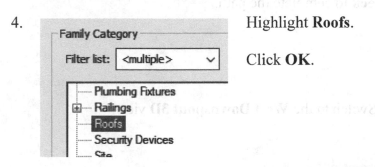 Highlight **Roofs**.

Click **OK**.

5.  Type **Downspout-End** in the Name field.

Click **OK**.

6. Select **Swept Blend** from the Create tab on the tab.

7. Select **Sketch Path**.

8. 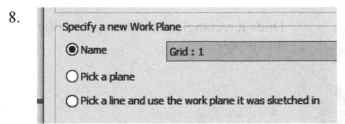 Enable **Name**.

Select **Grid: 1** from the drop-down list.

Click **OK**.

9. 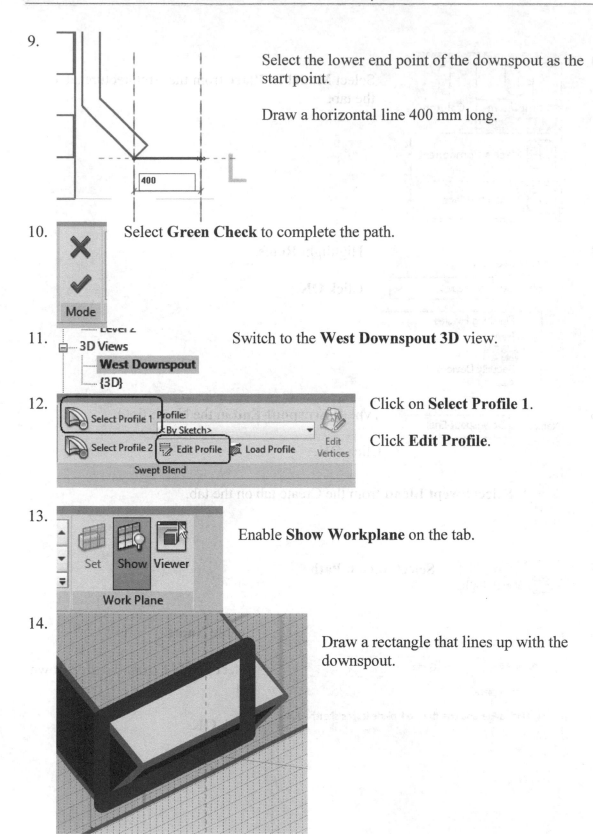 Select the lower end point of the downspout as the start point.

Draw a horizontal line 400 mm long.

400

10. Select **Green Check** to complete the path.

Mode

11. Switch to the **West Downspout 3D** view.

Level 2
3D Views
West Downspout
{3D}

12. Click on **Select Profile 1**.

Select Profile 1 Profile:
<By Sketch>

Click **Edit Profile**.

Select Profile 2 Edit Profile Load Profile Edit Vertices

Swept Blend

13. Enable **Show Workplane** on the tab.

Set Show Viewer

Work Plane

14. Draw a rectangle that lines up with the downspout.

15. 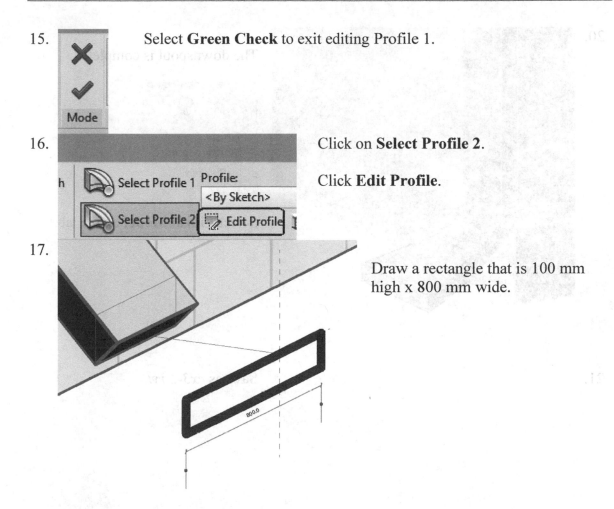 Select **Green Check** to exit editing Profile 1.

16. Click on **Select Profile 2**.

Click **Edit Profile**.

17. Draw a rectangle that is 100 mm high x 800 mm wide.

800.0

18. 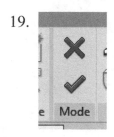 Select **Green Check** to exit editing Profile 2.

19. Select **Green Check** to complete the Swept Blend.

20.

The downspout is completed.

Select **Finish Model** on the tab.

21.

Save as *ex3-6.rvt*.

Exercise 3-7

Placing a Mass

Drawing Name: **new**
Estimated Time to Completion: 15 Minutes

Scope
Modifying a conceptual mass

Solution

1. Start a new project using the *Metric-Architectural* template.

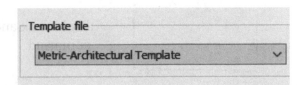

Template file

Metric-Architectural Template

2.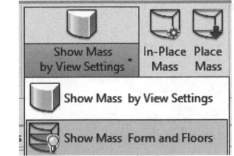

    Activate the Massing & Site tab.

    Enable **Show Mass Form and Floors** from the Conceptual Mass panel.

3.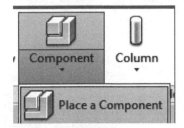

    Activate the Architecture tab.

    Select the **Component→Place a Component** tool on the Build panel.

4.

    Select **Load Family** from the Mode panel.

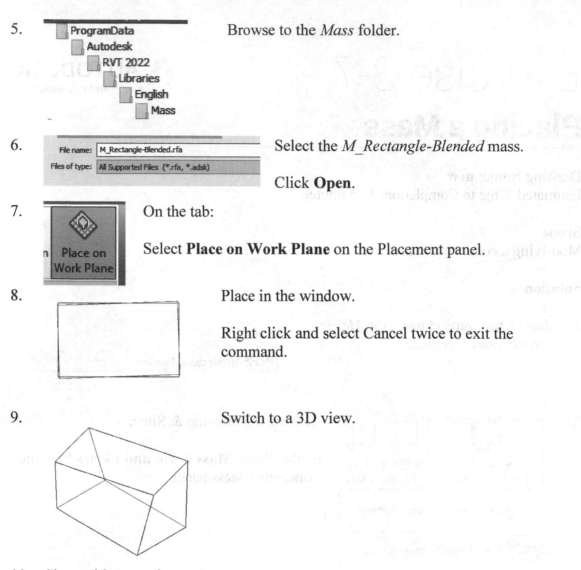

5.   Browse to the *Mass* folder.

6.   Select the *M_Rectangle-Blended* mass.

     Click **Open**.

7.   On the tab:

     Select **Place on Work Plane** on the Placement panel.

8.   Place in the window.

     Right click and select Cancel twice to exit the command.

9.   Switch to a 3D view.

10. Close without saving.

Exercise 3-8

Placing Mass Floors

Drawing Name: mass_floors.rvt
Estimated Time to Completion: 5 Minutes

Scope
Placing mass floors

Solution

1. Activate the Massing & Site tab.

   Enable **Show Mass Form and Floors** from the Conceptual Mass panel.

2. Select the mass so that it highlights.

   Select the **Mass Floors** tool from the tab.

3. Place a check on all the levels.

   Click **OK**.

   Mass Floors
   - ☑ Level 1
   - ☑ Level 2
   - ☑ Level 3
   - ☑ Level 4
   - ☑ Level 5

4. Select the mass floor for Level 4.
   Use TAB to help select.
   What is the floor area?

   You should see a floor area of 1485.80 SF.
   Close without saving.

Mass Floor (1)	
Dimensions	
Floor Perimeter	159' 2 57/64"
Floor Area	1485.80 SF
Exterior Surface Area	2302.81 SF
Floor Volume	7462.14 CF
Level	Level 4

Things to remember about Masses:

- The default material for a mass is 5 percent transparent.
- Masses will not print unless the category is enabled in Visibility/Graphics Overrides.
- Masses are created from a single closed profile.
- Masses are a nested entity. In order to modify the profile, you have to open the mass up for editing and then open the desired form component up for editing.
- Masses can be comprised of multiple forms, a combination of voids and solids.

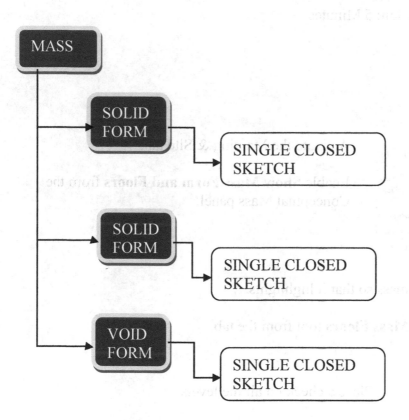

Room Calculation Point

In most instances, families placed within a room are associated with the room in a schedule. There are a few circumstances in which a family instance does not report its room correctly. When families such as furniture, doors, windows, casework, specialty equipment and generic models are placed in a project, sometimes parts of their geometry are located outside a room, space, or within another family, which results in no calculable values being reported.

Room Calculation Points are used to make loadable families "room aware." Once you enable the room calculation point, the family will display the correct room in the project schedule.

To enable and modify the Room Calculation Point to reorient room aware families:

1. Select the family instance in the drawing area.

2. Click Modify | <Element> tab ➤ Mode panel ➤ Edit Family.

3. In the Family Editor, open a floor plan view of the family.

4. On the Properties palette, select the Room Calculation Point parameter in the Other section. The point is now visible in the drawing area as a green dot.

5. Select the Room Calculation Point and move it to a location that will not be obscured by geometry when placed in the project.

6. Click Modify tab ➤ Family Editor panel ➤ Load into Project.

There will be at least one question on the Professional exam where you will be asked to adjust the room calculation point on a loadable family.

Exercise 3-9

Room Calculation Point

Drawing Name: **i_room_calculation_point**
Estimated Time to Completion: 20 Minutes

Scope

Modify a room calculation point on a family so it associates with the correct room in a schedule.

Solution

1.

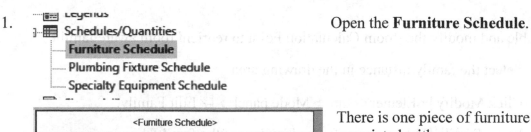

Open the **Furniture Schedule**.

There is one piece of furniture not associated with a room.

2.

Highlight the chair with no room shown.

3.

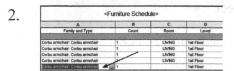

Select **Highlight in Model** on the tab.

4.

Revit ×

There is no open view that shows any of the highlighted elements. Searching through the closed views to find a good view could take a long time. Continue?

OK   Cancel

Click **OK**.

5.  Select the **Show** button until the view opens where you see the chair.

6. 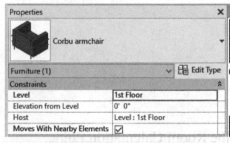 When the best view with the chair opens, Click **Close** on the dialog.

If you right click on the chair, you will see that Edit Family is grayed out.

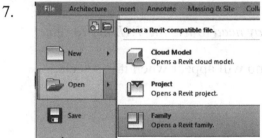 On the Properties pane, you see that the chair is a Corbu armchair.

7. 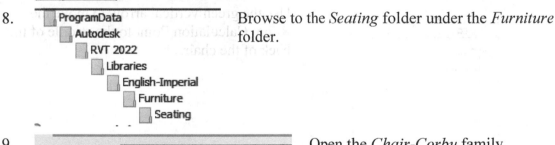 Go to the Application menu.
Select **Open→Family**.

8. 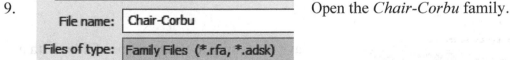 Browse to the *Seating* folder under the *Furniture* folder.

9. 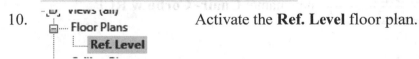 Open the *Chair-Corbu* family.

10. Activate the **Ref. Level** floor plan.

11.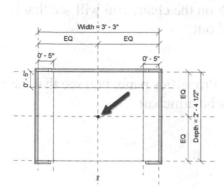

Place a check next to **Room Calculation Point**.

This makes the family "room aware."

12.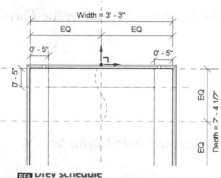

A green dot will appear to indicate the location of the Room Calculation Point.

13.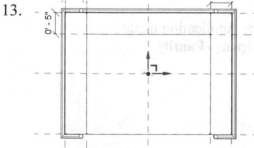

Select the Room Calculation Point.

You may need to use the TAB key to select.

A gizmo will appear when it is selected.

14.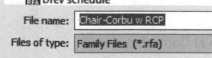

Use the green vertical arrow to move the Room Calculation Point to the middle of the back of the chair.

15.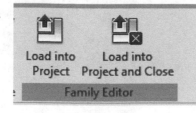

Save the family to your exercise folder with a new name: **Chair- Corbu w RCP**.

16.

Select the **Load into Project and Close** command from the tab.
This loads the family into the open project and closes the family file.

17.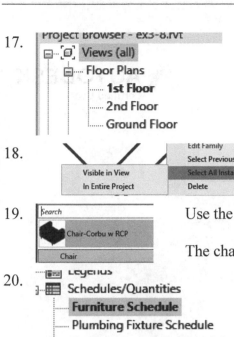

Cancel out of the command if your cursor shows you placing a chair.

Verify that the 1st Floor view is open and active.

18. Select the chair. Right click and **Select All Instances→Visible in View**.

19. Use the Type Selector to select the **Chair- Corbu w RCP.**

The chair will shift position based on the new RCP.

20. Open the **Furniture Schedule**.

21.

Note the chair is now associated with a room.

Close without saving.

<Furniture Schedule>

A	B	C	D
Family and Type	Count	Room	Level
Chair-Corbu w RCP: Chair	1	LIVING	1st Floor
Chair-Corbu w RCP: Chair	1	LIVING	1st Floor
Chair-Corbu w RCP: Chair	1	LIVING	1st Floor
Chair-Corbu w RCP: Chair	1	LIVING	1st Floor
Chair-Corbu w RCP: Chair	1	LIVING	1st Floor

Toposurfaces

A topographical surface (a toposurface) is created using points or imported data. You can create toposurfaces in 3D views or site plans.

When viewing a toposurface, consider the following:

- **Visibility**. You can control the visibility of topographic points. There are 2 topographic point subcategories: Boundary and Interior. Revit classifies points automatically.
- **Triangulation edges**. Triangulation edges for toposurfaces are turned off by default. You can turn them on by selecting them from the Model Categories/Topography category in the Visibility/Graphics dialog.

Exercise 3-10

Creating a Toposurface

Drawing Name: **C_Condo_complex.rvt**
Estimated Time to Completion: 40 Minutes

Scope
Create a toposurface.

Solution

1.   Floor Plans   Activate the **Site** view.
         Level 1   Turn off the visibility of grids, elevations, and sections.
         Level 2
         Level 3
         Roof
         Site

2.   Activate the **Massing & Site** tab.

     Toposurface   Massing & Site

     Select the **Toposurface** tool.

3.   Use the **Place Point** tool to create an outline of a lawn surface.

     Place Point

4. 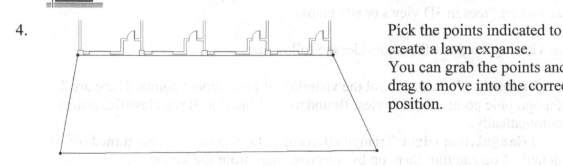   Pick the points indicated to create a lawn expanse.
You can grab the points and drag to move into the correct position.

5. Click on the **Modify** tool on the tab to exit out of the add points mode.

6. Left click in the **Material** column on the Properties pane.

7. In the search field, type site.
Site materials will then be listed.
Highlight **Site - Grass**.

8. On the Graphics tab,
enable **Use Render Appearance.**
Click **OK** to close the Material Browser dialog.

9. You should see **Site-Grass** in the Properties pane.

10. Select the **Green Check** on the Surface panel to **Finish Surface**.

11. Switch to **Realistic** to see the grass material.

12. Save as *ex3-10.rvt.*

Exercise 3-11

Creating a Building Pad

Drawing Name: **building_pad.rvt**
Estimated Time to Completion: 40 Minutes

Scope
Create a toposurface.
Create a building pad.

Solution

1. ⊟ Floor Plans
   — Level 1
   — Level 2
   — Level 3
   — Roof
   — **Site**

   Activate the **Site** view.
   Turn off the visibility of grids, elevations, and sections.

2. ⊞ Wireframe
   ⊟ Hidden Line
   ⊞ Shaded
   ⊞ Consistent Colors
   ⊞ Realistic
   ■ Ray Trace

   Switch the display to **Hidden Line** to make it easier to place the building pad.

3. Toposurface

   Massing & Site

   Activate the **Massing & Site** tab.

   Select the **Toposurface** tool.

4. Place Point

   Use the **Place Point** tool to create an outline of a lawn surface.

5. 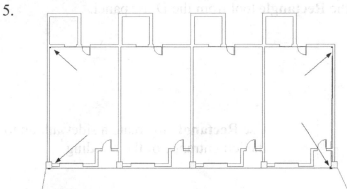 Select the four corners of the building to form a rough rectangle.

6.  Click on the **Modify** tool on the tab to exit out of the add points mode.

7. 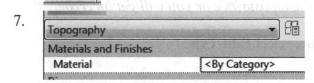 Left click in the **Material** column on the Properties pane.

8. In the search field, type **concrete**.
Highlight **Concrete - Cast In-Place Concrete**.

9. Enable **Use Render Appearance** in the Material Editor dialog.
Click **OK** to close the Material Browser dialog.

10. You should see **Concrete: Cast-In-Place Concrete** in the Properties pane. Click **OK**.

11. Select the **Green Check** on the Surface panel to **Finish Surface**.

12. Select the **Building Pad** tool from the Model Site panel.

Building Pad

13. Select the **Rectangle** tool from the Draw panel.

14. Use **Rectangle** to create a sidewalk up to the left entrance of the building.

The rectangle must be entirely within the toposurface or you will get an error message.

15. On the Properties pane, select **Edit Type**.

16. Select **Duplicate**.

17. Enter **Walkway** in the Name field.
Click **OK**.

18. Select **Edit** under Structure.

19. Assign the Masonry- Brick material to the structure layer.

20. Select Masonry-Brick as the material.

21. 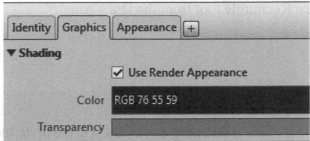 Enable **Use Render Appearance for Shading**.
Click **OK** to exit all the material dialogs.

22. In the Properties pane,
set the Height Offset from Level to **1"**.

This protrudes the walkway 1" above the toposurface so it appears better.

23. Select the **Green Check** on the Mode panel to **Finish Building Pad**.

24. Switch to a **3D** view.

25. 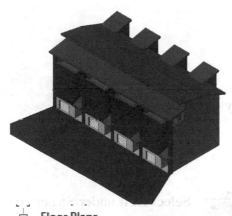 Use View Properties (VP) to set the style to **Shaded**.
Rotate the view so you can see the topo surface.

26.
```
Floor Plans
    Level 1
    Level 2
    Level 3
    Roof
    Site
```
Activate the **Site** view.

27. Select the **Building Pad** tool from the Model Site panel.

28. 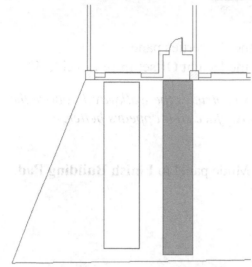 Select the **Rectangle** tool from the Draw panel.

29. Use **Rectangle** to create a drive up to the garage door of the building.

    The rectangle must be entirely within the toposurface or you will get an error message.

30. On the Properties pane, select **Edit Type**.

31. Select **Duplicate**.

32. Enter **Driveway** in the Name field. Click **OK**.

    Name: Driveway

33. Select **Edit** under Structure.

Parameter	Value
Construction	
Structure	Edit...
Thickness	1' 0"

34. Assign the **Site-Asphalt** material to the structure layer.

    asph

    Project Materials: All ▼

    Search results for "asph"

	Name
	Roofing - Asphalt Shingle
	Site - Asphalt

35.  Enable **Use Render Appearance** on the Graphics tab.
Click **OK**.

36.

	Function	Material
1	Core Boundary	Layers Above Wrap
2	Structure [1]	Site - Asphalt
3	Core Boundary	Layers Below Wrap

Verify that the correct material has been assigned.
Click **OK**.

37.

Pads	
Constraints	
Level	Level 1
Height Offset From Level	0' 1"
Room Bounding	☑

In the Properties pane,
set the Height Offset from Level to **1″**.

This protrudes the walkway 1″ above the toposurface so it appears better.

38. Select the **Green Check** on the Mode panel to **Finish Building Pad**.

39. 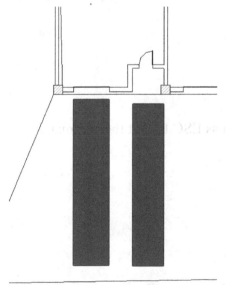 Hold down the CTL key.
Select the driveway and walkway building pads.

40. Select the **Copy** tool from the Modify panel.

41. ☑ Constrain ☐ Disjoin ☑ Multiple   Enable **Constrain**.
This enables ORTHO mode.
Enable **Multiple**.
This allows you to place more than one copy.

42.

Select a point to use as the basepoint for the Copy command.

I selected the upper corner of the doorway.

43.

Select the corresponding point on the next unit to place the copied elements.

44.

Press ESC to exit the command.

45. Save the file as *ex3-11.rvt*.

Exercise 3-12

Toposurface Properties

Drawing Name: **i_Bldg_Pad.rvt**
Estimated Time to Completion: 10 Minutes

Scope
Create a building pad.
Determine the volume of the building pad.

Solution

1.  Activate the **Site** floor plan.

2. Zoom into the outline shown. You will create a building pad using this outline.

3. Activate the Massing & Site tab.

   Select the **Building Pad** tool.

4. Select the **Pick** tool.

   Select the building pad outline.

5. 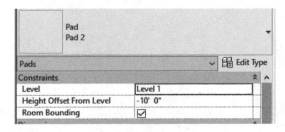 On the Properties pane:

   Set the Level to **Level 1**.

   Set the Height Offset from Level to **-10' 0"**.
   Select **Edit Type**.

6.

Parameter	Value
Construction	
Structure	Edit...
Thickness	1' 2"

Select **Edit Structure**.

7.

	Function	Material	Thickness
1	Core Boundary	Layers Above W	0' 0"
2	Structure [1]	<By Category>	1' 6"
3	Core Boundary	Layers Below Wr	0' 0"

Set the Thickness to **1' 6"**.

Click **OK**.

Close all the dialog boxes.

8.

Mode

Green check to complete the building pad.

9.

Dimensions	
Slope	
Perimeter	115' 1 225/256"
Area	752.91 SF
Volume	1129.35 CF

Look in the Properties pane.

The building pad should still be selected.

What is the volume of the pad you just created?

It should be 1129.35 CF.

Certified User Practice Exam

1. You can create a toposurface by placing points OR:
    A. creating from import
    B. drawing contour lines
    C. drawing a path
    D. picking points

2. To identify a family type of an element:
    A. Select the element, look on the Project Browser
    B. Hover over the element and read the description that is displayed
    C. Select the element, use Filter
    D. Use Insert, Load Family

3. True or False

    Furniture families are hosted.

4. True or False

    Window families are hosted.

5. To control the visibility of toposurfaces using the Visibility/Graphics dialog:

    A. Use Annotation Categories
    B. Use Model Categories
    C. Use Site->Topography under Model Categories
    D. Use Topography under Model Categories

6. To add a loadable family to a project, use the [fill in the blank] tab:

    A. Create
    B. Insert
    C. Family
    D. Architecture

Answers:
    1) A; 2) B; 3) F; 4) T; 5) D; 6) B

Certified Professional Practice Exam

1. To create a new system family you:

    A. Use a Revit family template
    B. Modify the instance parameters of an existing similar family
    C. Duplicate a similar system family and modify the type parameters
    D. Modify the type parameters of a similar system family

2. When a room schedule does not reflect the correct room for an element, the element family should be modified by:

    A. Moving it into the correct room
    B. Modifying the element's base point
    C. Making the element "room-aware"
    D. Modifying the room schedule

3. Room Calculation Points are used to:

    A. Calculate the square footage of a room
    B. Make loadable families "room aware"
    C. Calculate the volume of a room
    D. Create room schedules

4. Before you add a building pad, you need a:
    A. floor
    B. toposurface
    C. wall
    D. isolated foundation

5. To create a new door type of a door family without exiting the active project:
    A. Select the door, right click and select Edit Family
    B. Go to the Properties panel, select Edit Type
    C. Go to the Project Browser, locate the door family, right click and select New Type
    D. Go to the Insert tab, select Load Family

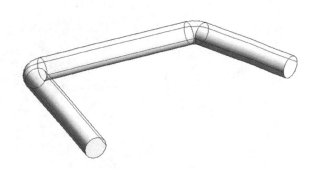

6. The mass shown was created using:

    A. Sweep
    B. Extrude
    C. Blend
    D. Swept Blend

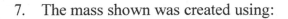

7. The mass shown was created using:

    A. Sweep
    B. Extrude
    C. Blend
    D. Swept Blend

Answers:
    1) C; 2) C; 3) B; 4) B; 5) C; 6) A; 7) C

Notes:

Managing Views

This lesson addresses the following exam questions:

- View Properties
- Object Visibility Settings
- Section Views
- Elevation Views
- 3D Views
- View Templates
- Scope Boxes
- Schedules
- Legends
- Rendering
- Revision Clouds

The Project Browser lists all the views available in the project. Any view can be dragged and dropped onto a sheet. Once a view is used or consumed on a sheet, it cannot be placed a second time on a sheet – even on a different sheet. Instead, you must create a duplicate view. You can create as many duplicate views as you like. Each duplicate view may have different annotations, line weight settings, detail levels, etc. Annotations are specific to a view. If a view is deleted, any annotations are also deleted.

Revit has bidirectional associativity. This means that changes in one view are automatically reflected in all views. For example, if you modify the dimensions or locations of a window in one view, the change is reflected in all the views, including the 3D view.

You can control the appearance of Revit elements using Object Visibility Settings. These settings control line color, line type, and line weight. You can create templates which have different Object Visibility Settings for different project types.

View Scale

The view scale controls the scale of the view as it appears on the drawing sheet.

You can assign a different scale to each view in a project. You can also create custom view scales.

Many view types in Revit contain a "View Scale" property- such a Floor Plans, Ceiling Plans, Sections, Elevations, Callouts, and Drafting Views. The "View Scale" parameter allows you to set the scale at which each particular view will be printed out. You can also modify the view scale to ensure that a view fits on a sheet.

Key Points

- Each view has its own "View Scale" property.
- You can change the scale of a view at any time- using the View Control Bar or View Properties.
- Revit maintains annotations (tags, dimensions, etc.) at their actual printed size, regardless of the scale of the view.

[image reference placement]

Exercise 4-1
Change the View Scale

Drawing Name: **view_scale.rvt**
Estimated Time to Completion: 10 Minutes

Scope
Changing a view scale.
Adding a custom view scale.

Solution

1. 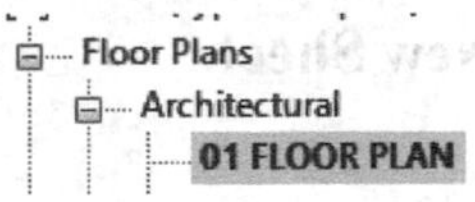 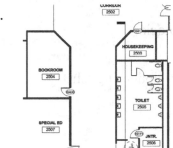 Activate the **01 FLOOR PLAN**.

2. Zoom into the area with Rooms 2504 and 2503.

3.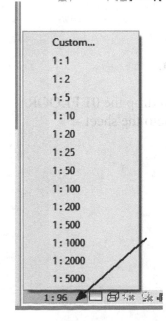

    Look down at the bottom left hand corner of the active view window. Locate the View Control Bar.

    The first button on the View Control Bar is the View Scale.

    You can see that this view is indeed set to **1:96**.

    Click on the View Scale button.

    A pop-up displays a list of all the available scales.

    At the top of the list is Custom. This is used to define a scale that is not available in the default list.

4. 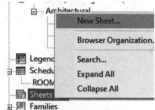 Change the view scale to **1:100**.

Did you notice how the room tags adjusted size in relation to the view scale?

5. 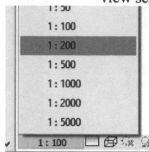 Change the view scale to **1:200**.

Did the room tags get bigger or smaller?

In fact, the room tags remained the same size. The model elements increased in scale.

6. In the browser, locate the Sheets category.

Highlight and right click to select **New Sheet**.

7. Click **Load** to load a new titleblock.

8. Browse to the *Titleblocks* folder under English.

9. Select *A2 metric.rfa*.
Click **Open**.
Click **OK** to create a new sheet.

10. 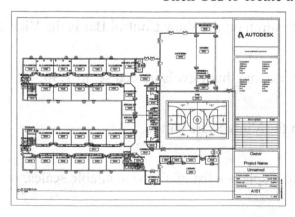 Drag and drop the 01 FLOOR PLAN onto the sheet.

11.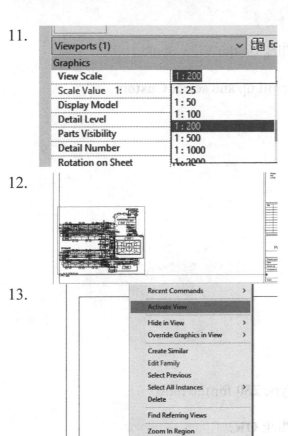

Select the view on the sheet.

On the Properties pane:

Set the View Scale to 1:500.

Click **Apply.**

12. The view updates.

13.

Select the view.
Right click and select **Activate View**.

The View Control bar is now visible.

14.

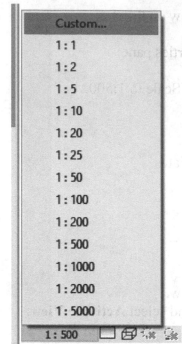

Click on the View Scale.

Scroll up and select **Custom**.

15.

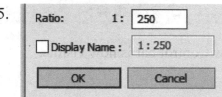

Type **250** for the ratio.

Click **OK**.

16.

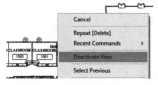

Right click on the view.

Select **Deactivate View**.

17.

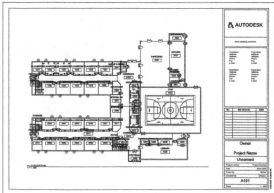

The view can now be adjusted to place on the sheet.

Close without saving.

Detail Level

Detail level determines the visibility of elements at different levels of detail.

For example, at the coarse level of detail, the different layers in a wall structure are not visible, just the outline of the wall. At the fine level of detail, all the layers in the wall are displayed.

You can set the detail level for newly created views based on a view scale. View scales are organized under the detail level headings: Coarse, Medium or Fine.

When you create a new view in your project and set its view scale, its detail level is automatically set according to the arrangement in the table.

By pre-defining detail levels, you can control the display of the same geometry at different view scales.

Exercise 4-2

Change the Detail Level of a View

Drawing Name: **detail_level.rvt**
Estimated Time to Completion: 15 Minutes

Scope
Defining Detail Levels for View Scales

Solution

1.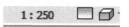

   Floor Plans
   Architectural
   01 FLOOR PLAN

   Activate the **01 Floor Plan**.

2. 1 : 250 ⬜ 🔲

   On the View Control bar:
   The View Scale is set to **1:250**.
   The Detail Level is set to **Coarse**.
   The Display Style is set to **Hidden Line**.

3.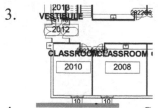

   You should be able to identify the different icons associated with Detail Level and Display Style for the certification exam.
   Zoom into the lower left area of the floor plan – where Rooms 2010 and 2008 are located.

4. Additional Settings

   Switch to the Manage tab on the ribbon.

   Select **Additional Settings** on the ribbon.

5.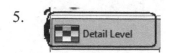

   Select **Detail Level** from the drop-down list.

6.

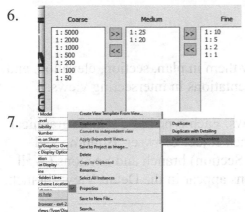

Note that each View Scale has been assigned a Detail Level setting.

Click **OK**.

7.

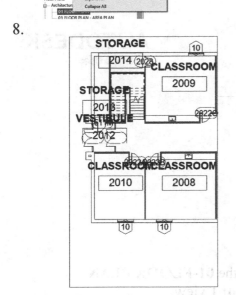

Highlight the **01 FLOOR PLAN**.

Right click and select **Duplicate View→Duplicate as a Dependent.**

8.

Crop the view using the crop region so only Rooms 2008, 2009, and 2010 are displayed.

Change the view scale to **1:10**.

The Detail Level is set to **Coarse**.
The View Display is set to **Hidden Line**.

You may recall on the Detail Level settings, the Detail Level for a View Scale of 1:10 should be Fine.

9.

Set the Detail Level to **Fine.**

Inspect the wall.

10.

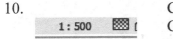

Change the View Scale to **1:500**.
Change the View Scale to **Coarse**.

Notice how the wall display changes.

11.

Close without saving.

Section Views

Sections views cut through the model. You can draw them in plan, section, elevation, and detail views. Section views display as section representations in intersecting views.

You can create building, wall, and detail section views. Each type has a unique graphical display, and each is listed in a different location in the Project Browser. Building and wall section views display in the Sections (Building Section) branch and Sections (Wall section) branch of the Project Browser. Detail sections appear in the Detail Views branch.

Exercise 4-3

Create a Section View

Drawing Name: **section_view.rvt**
Estimated Time to Completion: 5 Minutes

Scope
Create a section view.

Solution

1.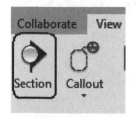
    Activate the **01-FLOOR PLAN – Dependent 1** view.

2.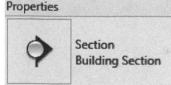
    Switch to the View tab.

    Select the **Section** tool.

3. Properties

    ◆ Section
    Building Section

    On the Properties pane:

    Verify the **Building Section** is selected.

4.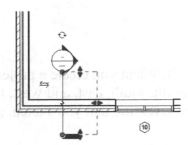

Left click inside the building to place the bubble.
Drag the mouse down.
Left click to place the section tail.

Adjust the section line so the section depth contains only the wall.

5.

Section
Building Section

Views (1)

Graphics

View Scale	1 : 10
Scale Value 1:	10
Display Model	Normal
Detail Level	Fine
Parts Visibility	Show Original

With the section line selected:

Go to the Properties pane.
Change the View Scale to **1:10**.
Change the Detail Level to **Fine**.

6.

Identity Data	
View Template	<None>
View Name	South Wall
Dependency	Independent
Title on Sheet	
Referencing Sheet	A101
Referencing Detail	1

Properties help   Apply

Scroll down.
Change the View Name to **South Wall**.

Click **Apply**.

7.

- Sections (Building Section)
  - Architectural
    - Section 1
    - Section 2
    - Section 3
    - **South Wall**

Open the **South Wall** section view.

8.

1 : 10

Verify the View Scale is set to **1:10**.
Verify the Detail Level is set to **Fine**.

9.

Close without saving.

Elevation Views

Elevation views are part of the default template in Revit. When you create a project with the default template, 4 elevation views are included: north, south, east, and west. It is in elevation views where you sketch level lines. For each level line that you sketch, a corresponding plan view is created.

You can create additional exterior elevation views and interior elevation views. Interior elevation views depict detailed views of interior walls and show how the features of the wall should be built. Examples of rooms that might be shown in an interior elevation are kitchens and bathrooms.

Each elevation has type properties for elevation tags, callout tags, and reference labels. To define the look of elevation tags and callout tags, use the Manage tab ➤ Settings panel ➤ Advanced Settings drop-down ➤ ⌑ (Callout Tags) or ⌂ (Elevation Tags). The Reference Label parameter sets the text displayed next to the elevation tag when the elevation is a reference elevation.

Exercise 4-4

Create an Elevation View

Drawing Name: **elevation.rvt**
Estimated Time to Completion: 5 Minutes

Scope
Define elevation tag naming.
Create an elevation view.

Solution

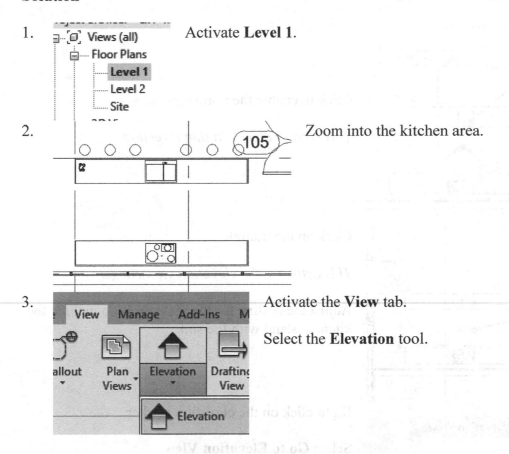

1. Activate **Level 1**.

2. Zoom into the kitchen area.

3. Activate the **View** tab.

   Select the **Elevation** tool.

4.

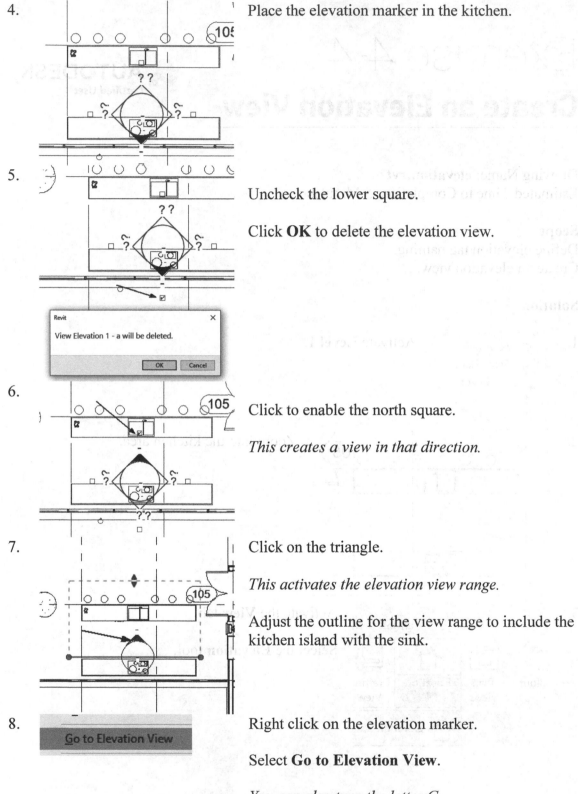

Place the elevation marker in the kitchen.

5.

Uncheck the lower square.

Click **OK** to delete the elevation view.

Revit — View Elevation 1 - a will be deleted.

6.

Click to enable the north square.

This creates a view in that direction.

7.

Click on the triangle.

This activates the elevation view range.

Adjust the outline for the view range to include the kitchen island with the sink.

8.

Go to Elevation View

Right click on the elevation marker.

Select **Go to Elevation View**.

You can also type the letter G.

9.

Set the view display to **Realistic Colors**.

1 : 1

10. 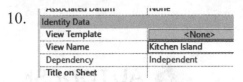 Change the view name to **Kitchen Island** on the Properties pane.

11. Close without saving.

3D Views

There are two kinds of 3D views: orthographic and perspective. *Orthographic 3D views* show the building model in a 3D view where all components are the same size, regardless of the camera's distance. Orthographic 3D views are also referred to as parallel 3D views because the elements are projected onto a viewing plane, which in turn make the extents parallel. *Perspective 3D views* on the other hand, show the building model in a 3D view where components that are farther away appear smaller, and components that are closer appear larger. This simulates the way human vision sees the world.

The default 3D view, which is easily accessible in Revit, is an orthographic 3D view. The default 3D view appears in the **Project Browser** under the **3D Views** category. The view is named **{3D}**, and the name is in curly brackets. When you see this view in the **Project Browser**, you know that it is the default 3D view.

You can create a 3D view using two methods.

In Method 1, simply duplicate the default 3D view, then modify it by orbiting the model, zooming in or out or using a section box. Then lock the display.

In Method 2, use a 3D camera to create a 3D view. The 3D camera tool allows you to create a perspective view and focus on a specific area of the model.

Exercise 4-5
Create a 3D View

Drawing Name: **3D_View.rvt**
Estimated Time to Completion: 15 Minutes

Scope
Create a 3D view
Orient to a View
Lock and Unlock a 3D View
3D Camera
Navigation Wheel

Solution

1. Activate the **Kitchen** 3D View.

   - 3D Views
     - Approach
     - From Yard
     - **Kitchen**
     - Living Room

2.  Toggle to unlock the view on the View Control bar.

3. 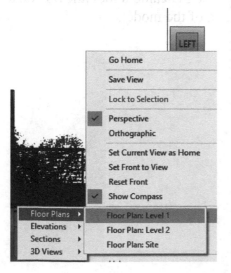 Place the mouse over the Viewcube.
   Right click and select **Orient to a View→Floor Plans→Floor Plan: Level 1**.

4.

The view re-orients to show a floor plan view of the kitchen area.

5.

Hold down the SHIFT key and the middle mouse button and the view will orbit.

You are still in a 3D view; it was just re-oriented to the plan view.

6.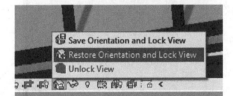

On the View Control bar:

Select **Restore Orientation and Lock View**.

7.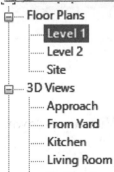

The original view is displayed.

8.

Floor Plans
  Level 1
  Level 2
  Site
3D Views
  Approach
  From Yard
  Kitchen
  Living Room

Activate the Level 1 floor plan.

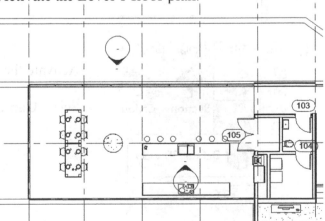

9.

Highlight the Kitchen 3D view.
Right click and select **Show Camera**.

The camera used to define the Kitchen 3D view is displayed.

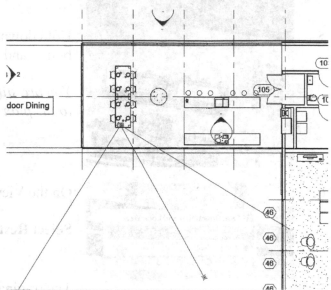

10.

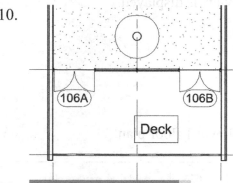

Zoom into the deck area.

11.

Activate the View tab.

Select **Camera** from the 3D View drop-down.

12.

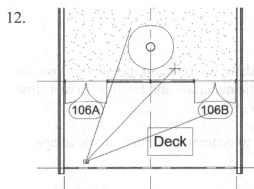

Place a camera in the lower left of the deck area.

Point the camera at Door 106B.

Left click to place.

13.

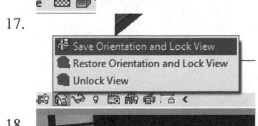

Change the View Name to **Deck** in the Properties pane.

14.

Launch the Navigation Wheel from the Navigation bar on the right side of the screen.

15.

Use ORBIT, PAN, and LOOK to adjust the camera view.

16. Change the Display Style to **Realistic**.

17.

Save Orientation and Lock View on the View Control bar

18.

Your view should look similar to this.

Close without saving.

Rendering

Rendering is used to present a design to clients or share it with team members. You can use the Realistic visual style, which displays realistic materials and textures in real-time or render the model to create a photorealistic image.

The rendering engine uses complicated algorithms to generate a photorealistic image from a 3D view of a building model.

The amount of time required to create the rendered image varies depending on many factors, such as the numbers of model elements and artificial lights, the complexity of the materials, and the size or resolution of the image. Reflections, refractions, and soft shadows can also increase render time.

Render performance is a balance between the quality of the resulting image and the resources (time, computing power) that can be devoted to the effort. Low quality images are generally quick to produce, while high quality images can require significantly more time.

Before rendering an image, consider whether you need a high-quality image or a draft quality image. In general, start by rendering a draft quality image to see the results of the initial settings. Then refine materials, lights, and other settings to improve the image. As you get closer to the desired result, you can use the medium quality setting to produce a more realistic image. Use the high-quality setting to produce a final image only when you are sure that the material render appearances and the render settings will give the desired result.

Exercise 4-6

Render a View

Drawing Name: **render.rvt**
Estimated Time to Completion: 15 Minutes

Scope
Create a rendering.
Control rendering options.
Save a rendering to the project.

Solution

1.   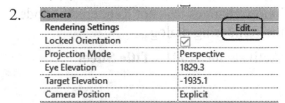   Activate the **3D Deck** view.

2.   On the Properties pane,
     under Camera,
     select **Edit** for Rendering Settings.

Camera	
Rendering Settings	Edit...
Locked Orientation	☑
Projection Mode	Perspective
Eye Elevation	1829.3
Target Elevation	-1935.1
Camera Position	Explicit

3.      Set the Quality Setting to **Medium**.
     Click **OK**.

4.   Select the **Rendering** tool located on the Display
     Control bar.

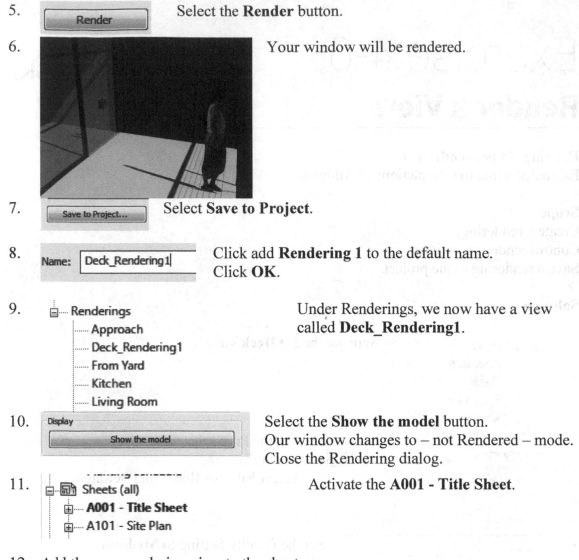

5.   Render   Select the **Render** button.

6.   Your window will be rendered.

7.   Save to Project...   Select **Save to Project**.

8.   Name: Deck_Rendering1   Click add **Rendering 1** to the default name.
                          Click **OK**.

9.   Renderings   Under Renderings, we now have a view
       Approach   called **Deck_Rendering1**.
       Deck_Rendering1
       From Yard
       Kitchen
       Living Room

10. Display   Select the **Show the model** button.
     Show the model   Our window changes to – not Rendered – mode.
                        Close the Rendering dialog.

11. Sheets (all)   Activate the **A001 - Title Sheet**.
       A001 - Title Sheet
       A101 - Site Plan

12. Add the new rendering view to the sheet.

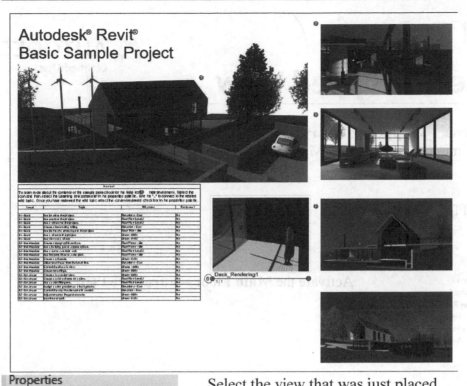

13. Select the view that was just placed.

On the Properties panel:

Select **Viewport: No Title** using the Type Selector.

14. Save the file as *ex4-6.rvt*.

Exercise 4-7
Create a Cropped View

Drawing Name: **i_cropped_views.rvt**
Estimated Time to Completion: 5 Minutes

Scope
Create a cropped view.

Solution

1.   Floor Plans
                Ground Floor
                Lower Roof
                Main Floor
                Main Floor- Furniture Plan
                Main Roof
                Site
                T. O. Footing
                T. O. Parapet

    Activate the **Main Floor – Furniture Plan** view.

2. | Referencing Detail | |
    |---|---|
    | **Extents** | |
    | Crop View | ☑ |
    | Crop Region Visible | ☑ |
    | Annotation Crop | ☐ |

    In the Properties Pane:
    Scroll down the window and
    Enable **Crop View**.
    Enable **Crop Region Visible**.

3.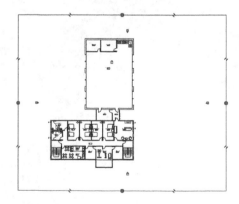

    Zoom out.

    Select the viewport rectangle.

4.

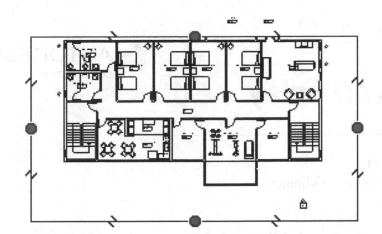

Use the bubbles to position the viewport so that only the furniture floor plan is visible.

5. Select **Hide Crop Region** using the tool in the View Control bar. Click **OK**.

6. Close without saving.

Exercise 4-8
Change View Display

Drawing Name: **i_views.rvt**
Estimated Time to Completion: 10 Minutes

Scope
Use Temporary Hide/Isolate to control visibility of elements.
Change Line Width Display.
Change Object Display Settings.

Solution

1.
Floor Plans
— Ground Floor
— Lower Roof
— **Main Floor**
— Main Floor- Furniture Plan
— Main Roof
— Site

Activate the **Main Floor** floor plan.

2.
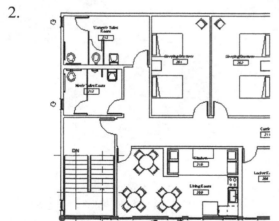

Select one of the exterior walls so it is highlighted.

3.
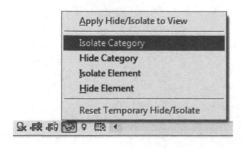

Select the **Temporary Hide/Isolate** tool.
Right click and select **Isolate Category**.

4. 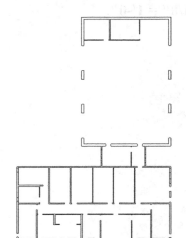 Only the exterior walls are visible.

5. 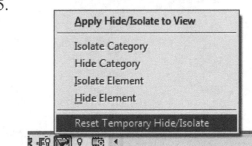 Select the **Temporary Hide/Isolate** tool. Right click and select **Reset Temporary Hide/Isolate**.

This restores the view.

6. 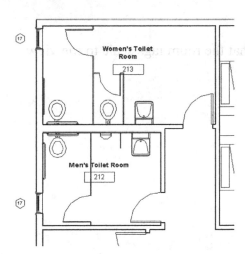 Zoom into the region where the lavatories are located.

7.

Custom...
12" = 1'-0"
6" = 1'-0"
3" = 1'-0"
1 1/2" = 1'-0"
1" = 1'-0"
3/4" = 1'-0"
1/2" = 1'-0"
3/8" = 1'-0"
1/4" = 1'-0"
3/16" = 1'-0"
1/8" = 1'-0"
1" = 10'-0"
3/32" = 1'-0"
1/16" = 1'-0"
1" = 20'-0"
3/64" = 1'-0"
1" = 30'-0"
1/32" = 1'-0"
1" = 40'-0"
1" = 50'-0"
1" = 60'-0"
1/64" = 1'-0"
1" = 80'-0"
1" = 100'-0"
1" = 160'-0"
1" = 200'-0"
1" = 300'-0"
1" = 400'-0"
1/4" = 1'-0"

Change the view scale to **1/8″ = 1′-0″**.

8. Note that the room tags scale to the view.

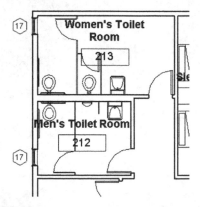

9. **Modify**   Activate the **Modify** ribbon.

10. Select the **Linework** tool on the View panel.

11. Set the Line Style to **Overhead**.

12. 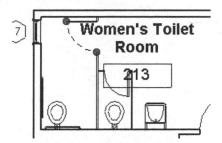 Select the door swing on the toilet cubicle.

Note that the door swing's appearance changes.

13. Activate **Section 1**.

14. Activate the Manage ribbon.

Select **Settings→Object Styles**.

15.

Detail Items	1		■ Black
Doors	2	2	■ Black
Elevation Swing	1	1	■ Black
Frame/Mullion	3	3	■ Magenta
Glass	1	4	■ Black
Hidden Lines	2	2	■ Blue
Opening	1	3	■ Black
Panel	3	5	■ Blue
Plan Swing	1	1	■ Black

Expand the **Doors** category on the Model Objects tab.
Change the Line Weight, Line Color, and Line Pattern for the Panel and Frame.
Click **Apply** to see the changes.

16. 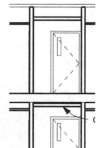 You can move the dialog over so you can see how the display is changed.

Click OK to close the dialog.

Note that linework changes are specific to the view, but object settings changes affect all views.

17. Close without saving.

Exercise 4-9

Duplicating Views

Drawing Name: **duplicating_views.rvt**
Estimated Time to Completion: 15 Minutes

Scope
Duplicate view with Detailing
Duplicate view as Dependent
Duplicate view
Understand the difference between the different duplicating views options

Solution

1.

    Floor Plans
       Level 1
       Level 2
       Level 3
       Site

Activate the **Level 1** floor plan.

Note that the doors all have door tags.

2.

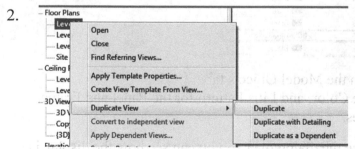

Highlight Level 1.
Select **Duplicate View→Duplicate**.

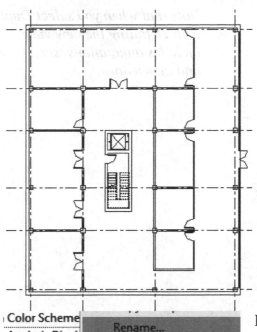

When you select Duplicate View→Duplicate, any annotations, such as tags and dimensions, are not duplicated.

3.   Highlight the copied level 1.

Right click and select **Rename**.

Name: Level 1 - No Annotations

Type **Level 1- No Annotations**.

Click **OK**.

4.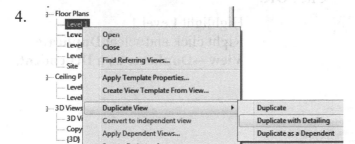

Highlight Level 1.
Right click and select **Duplicate View→Duplicate with Detailing**.

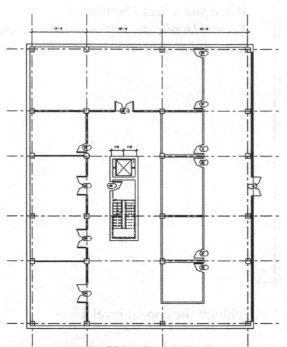

Note that when you select Duplicate with Detailing the new view includes annotations, such as tags and dimensions.

5.

Highlight the copied level 1.

Right click and select **Rename**.

Name: Level 1-Original Annotations

Type **Level 1- Original Annotations**.

Click **OK**.

6.

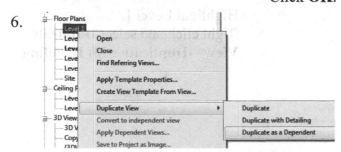

Highlight Level 1.
Right click and select **Duplicate View→Duplicate as a Dependent**.

7.
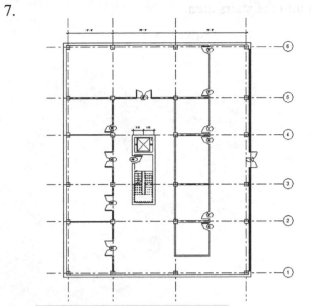

Notice that the duplicated view includes the annotations.

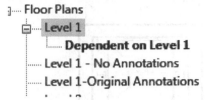

In the Project Browser, the dependent view is listed underneath the parent view.

8. Rename the dependent view **Level 1- Stairs area**.

Name: Level 1 - Stairs Area

9. Activate **Level 1- Stairs area**.

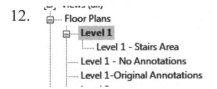

10. In the Properties pane,
Enable **Crop View**.
Enable **Crop Region Visible**.

Extents	
Crop View	☑
Crop Region Vis...	☑
Annotation Crop	☑

11. Adjust the crop region to focus the view on the stairs area.

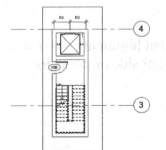

12. Activate **Level 1**.

13. 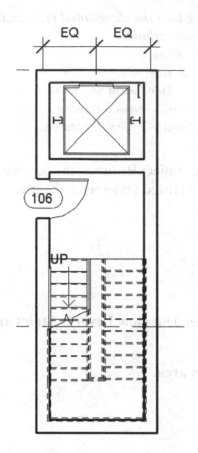 Zoom into the stairs area.

14. **Annotate**   Activate the **Annotate** ribbon.

15. **Tread Number**   Select the **Tread Number** tool on the Tag panel.

Tread Numbers can only be applied to Component-based stairs.

16. Select the middle line that highlights when your mouse hovers over the left side of the stairs.

17. 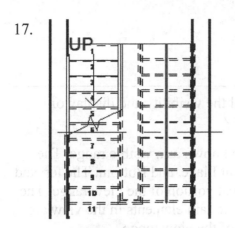 Select the middle line on the right side of the stairs. Right click and select CANCEL to exit the command.

18. 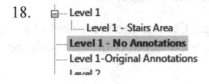 Activate **Level 1- Original Annotations**.

Notice that the new annotations - the tread numbers - are not visible in this view.

19. 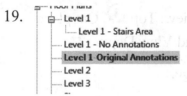 Activate **Level 1 - No Annotations**.

Notice that the new annotations - the tread numbers - are not visible in this view.

20. 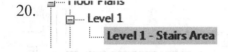 Activate the **Level 1- Stairs area** dependent view.

Notice that any annotations added to the parent view are added to the dependent view.

21. Close without saving.

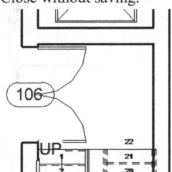

Extra: Change the door labeled 106 to a double flush door. Which Level 1 views display the new door type?

View Range

The view range is a set of horizontal planes that control the visibility and display of objects in a plan view.

Every plan view has a property called **view range**, also known as a visible range. The horizontal planes that define the view range are Top, Cut Plane, and Bottom. The top and bottom clip planes represent the topmost and bottommost portion of the view range. The cut plane is a plane that determines the height at which certain elements in the view are shown as cut. These 3 planes define the primary range of the view range.

View depth is an additional plane beyond the primary range. Change the view depth to show elements below the bottom clip plane. By default, the view depth coincides with the bottom clip plane.

The following elevation shows the view range ⑦ of a plan view: Top ①, Cut plane ②, Bottom ③, Offset (from bottom) ④, Primary Range ⑤, and View Depth ⑥.

The plan view on the right shows the result for this view range.

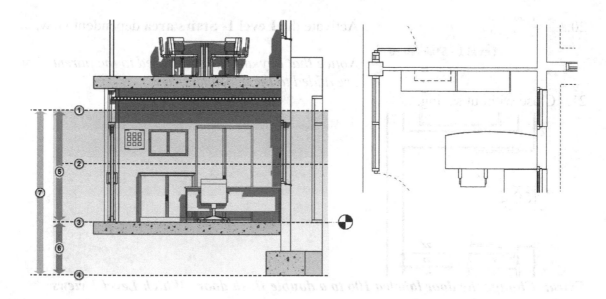

You will have at least one question on the Professional exam where you need to answer a question regarding one of the settings in the dialog.

Exercise 4-10

View Range

Drawing Name: **i_view_range.rvt**
Estimated Time to Completion: 5 Minutes

Scope
Determine the view depth of a view.

Solution

1.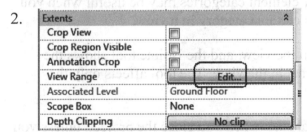
   Floor Plans
   ── Ground Floor
   ── Lower Roof
   ── Main Floor
   ── Main Roof
   ── Site
   ── T.O. Footing
   ── T.O. Parapet

   Activate the **Site** view.

2.

Extents	☆
Crop View	☐
Crop Region Visible	☐
Annotation Crop	☐
View Range	Edit...
Associated Level	Ground Floor
Scope Box	None
Depth Clipping	No clip

   In the Properties pane:
   Scroll down to **View Range** located under the Extents category.
   Select the **Edit** button.

3. << Show   Select the **Show** button located at the bottom left of the dialog.

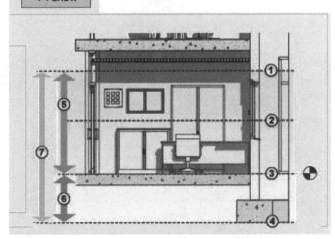

A representation of how view range settings work is displayed to assist you in defining the view range of a view.

Each number represents a field in the dialog box.

The elevation shows the view range ⑦ of a plan view.

4.

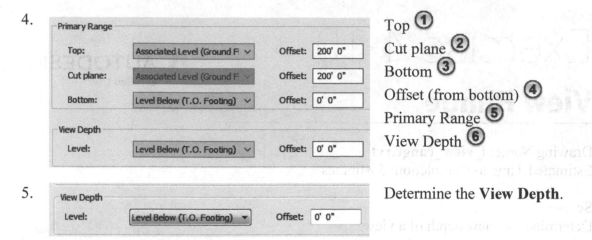

Top ①
Cut plane ②
Bottom ③
Offset (from bottom) ④
Primary Range ⑤
View Depth ⑥

5.

Determine the **View Depth**.

6. Click **OK**.

7. Close without saving.

Hidden Elements

Temporarily hiding or isolating elements or element categories may be useful when you want to see or edit only a few elements of a certain category in a view.

The Hide tool hides the selected elements in the view, and the Isolate tool shows the selected elements and hides all other elements in the view. The tool affects only the active view in the drawing area.

Element visibility reverts back to its original state when you close the project, unless you make the changes permanent. Temporary Hide/Isolate also does not affect printing.

Exercise 4-11

Reveal Hidden Elements

Drawing Name: **i_visibility.rvt**
Estimated Time to Completion: 5 Minutes

Scope
Turn on the display of hidden elements

Solution

1. 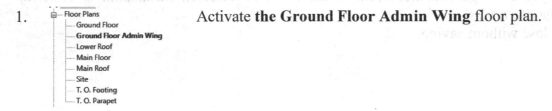   Activate **the Ground Floor Admin Wing** floor plan.

2. 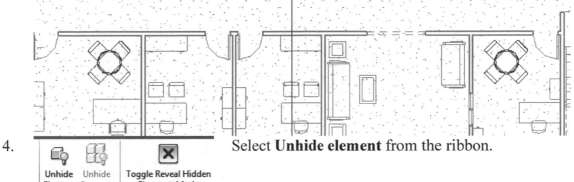   Select the **Reveal Hidden Elements** tool.

3. Items highlighted in magenta are hidden. Window around the two tables while holding down the CONTROL key to select them.

4.    Select **Unhide element** from the ribbon.

The tables will no longer be displayed as magenta (hidden elements).

5. 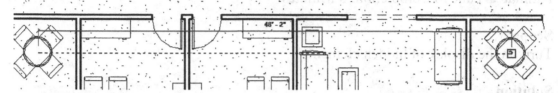 Select the **Close Hidden Elements** tool.

6. The view will be restored.

7. Select the **Measure** tool from the Quick Access toolbar.

8. Determine the distance between the center of the two tables.
   Did you get 48' 2"?

If you didn't get that measurement, check that you selected the midpoint or center of the two tables.

9. Close without saving.

View Templates

A view template is a collection of view properties, such as view scale, discipline, detail level, and visibility settings.

Use view templates to apply standard settings to views. View templates can help to ensure adherence to office standards and achieve consistency across construction document sets.

Before creating view templates, first think about how you use views. For each type of view (floor plan, elevation, section, 3D view, and so on), what styles do you use? For example, an architect may use many styles of floor plan views, such as power and signal, partition, demolition, furniture, and enlarged.

You can create a view template for each style to control settings for the visibility/graphics overrides of categories, view scales, detail levels, graphic display options, and more.

Filters

Filters provide a way to override the graphic display and control the visibility of elements that share common properties in a view.

For example, if you need to change the line style and color for different conduit types, you can create a filter that selects all conduits in the view that have the color 'red' in the description parameter. You can then select the filter, define the visibility and graphic display settings (such as line style and color), and apply the filter to the view. When you do this, all conduits that meet the criteria defined in the filter update with the appropriate visibility and graphics settings. You need to set up the view filters and then apply those filters to each view in order to display the conduits with the correct colors and linetypes.

Exercise 4-12
Create a View Template

AUTODESK.
Certified Professional

Drawing Name: **view_templates.rvt**
Estimated Time to Completion: 30 Minutes

Scope
Apply a wall tag.
Create a view template.
Create a view filter.
Apply view settings to a view.

Solution

1. Floor Plans
    Level 1
    Level 2
    Level 3
    Site
        Activate Level 1.

2. Annotate   Activate the **Annotate** ribbon.

3. Tag All   Select **Tag All**.

4. | Structural Framing Tags | Structural Framing Tag : Standard |
    |---|---|
    | Wall Tags | Wall Tag : 1/2" |
    | Wall Tags | Wall Tag : 1/4" |
    | Wall Tags | wall tag- fire rating |
    | Window Tags | Window Tag |

    Highlight the **Wall tag - fire rating** as the tag to be used and click **OK**.

    The wall tag - fire rating is a custom family. It was pre-loaded into this exercise but is included with the exercise files on the publisher's website for your use.

5. Zoom in to inspect the tags.

    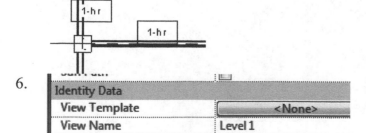

6. | Identity Data | |
    |---|---|
    | View Template | <None> |
    | View Name | Level 1 |

    In the Properties pane, click on the **<None>** button next to View Template.

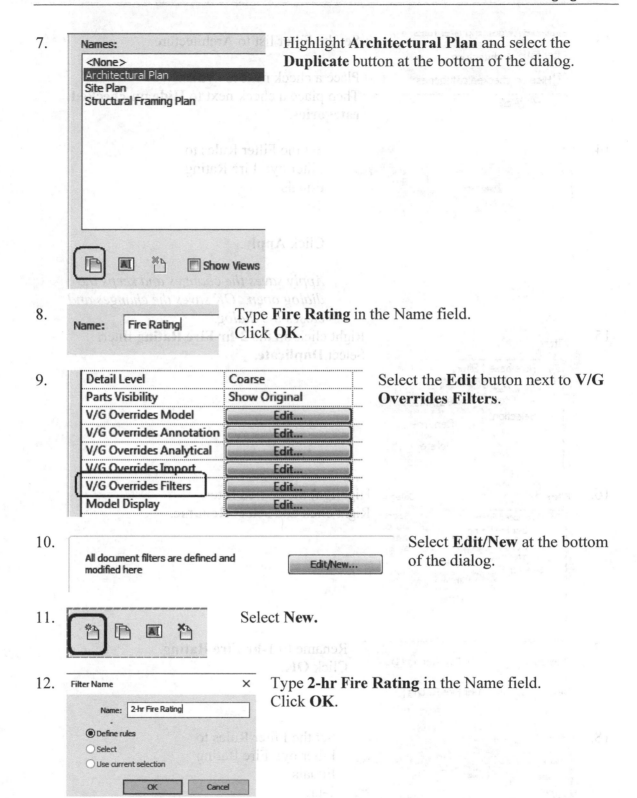

7.   Highlight **Architectural Plan** and select the **Duplicate** button at the bottom of the dialog.

8.   Type **Fire Rating** in the Name field. Click **OK**.

9.   Select the **Edit** button next to **V/G Overrides Filters**.

10.   Select **Edit/New** at the bottom of the dialog.

11.   Select **New.**

12.   Type **2-hr Fire Rating** in the Name field. Click **OK**.

13.

Set the Filter list to Architecture.

Place a check next to **Walls**.
Then place a check next to **Hide un-checked categories**.

14.

Set the Filter Rules to
Filter by: Fire Rating
Equals
2-hr.

Click **Apply**.

Apply saves the changes and keeps the dialog open. OK saves the changes and closes the dialog.

15.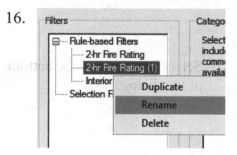

Right click on the **2-hr Fire Rating** filter.
Select **Duplicate**.

16.

Highlight the copied filter.
Right click and select **Rename**.

17.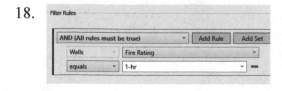

Rename to **1-hr Fire Rating**.
Click **OK**.

18.

Set the Filter Rules to
Filter by: Fire Rating
Equals
1-hr.

Click **Apply**.
Click **OK**.

19.

Name	Vi

No filters have been applied to this vi

Add   Remove

Select the **Add** button at the bottom of the Filters tab.

20.

Select one or more filters to insert.

- Rule-based Filters
  - 1-hr Fire Rating
  - 2-hr Fire Rating
  - Interior
- Selection Filters

Hold down the Control key.
Highlight the 1-hr and 2-hr fire rating filters and click **OK**.

21.

| Model Categories | Annotation Categories | Analytical Model Categories | Imported Categories | Filters |

Name	Visibility	Projection/Surface			Cut		Halftone
		Lines	Patterns	Transparen...	Lines	Patterns	
2-hr Fire Rating	☑	Override...	Override...	Override...	Override...	Override...	☐
1-hr Fire Rating	☑						☐

You should see the two fire rating filters listed.

If the interior filter was accidentally added, simply highlight it and select Remove to delete it.

22.

Name	Visibility	Projection/Surface		
		Lines	Patterns	Tra
2-hr Fire Rating	☑	Override...	Override...	C
1-hr Fire Rating	☑			

Highlight the **2-hr Fire Rating** filter.

23. Select the **Pattern Override** under Projection/Surface.

24.

Pattern Overrides
Foreground ☑ Visible
Pattern: 2 Hour

Select the **2 Hour** fill pattern for the Foreground.

25.

Pattern Overrides
Foreground ☑ Visible
Pattern: 2 Hour
Color: Blue

Set the Color to **Blue.**
Click **OK.**

26.

Name	Visibility	Projection/Surface		
		Lines	Patterns	T:
2-hr Fire Rating	☑		‖‖‖‖‖‖	
1-hr Fire Rating	☑	Override...	Override...	

Highlight the **1-hr Fire Rating** filter.

27. Select the **Pattern Override** under Projection/Surface.

28.

Pattern Overrides
Foreground ☑ Visible
Pattern: 1 Hour
Color: Magenta

Set the fill pattern for the Foreground to **1 Hour** and the Color to **Magenta.**
Click **OK.**

29. Apply the same settings to the Cut Overrides. Click **OK.**

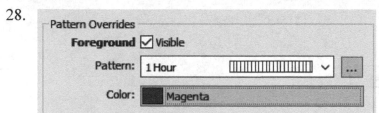

Name	Visibility	Projection/Surface			Cut	
		Lines	Patterns	Transparency	Lines	Patterns
2-hr Fire Rating	☑		‖‖‖‖‖‖			‖‖‖‖‖‖
1-hr Fire Rating	☑	Override...	‖‖‖‖‖	Override...	Override...	‖‖‖‖‖ Hidden

30.

Names:
<None>
Architectural Plan
Fire Rating
Site Plan
Structural Framing Plan

Highlight the **Fire Rating** View Template.

Click **OK.**

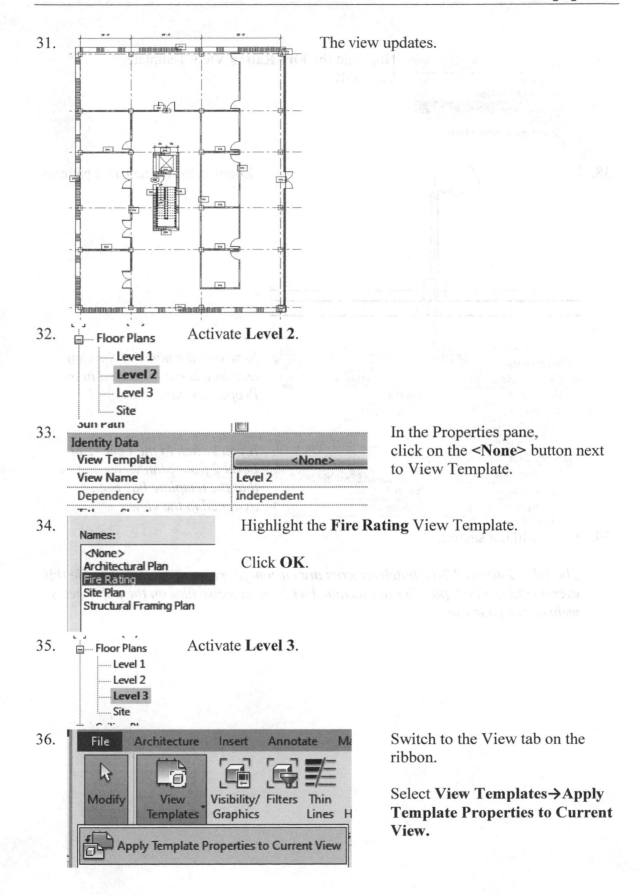

31. The view updates.

32. Activate **Level 2**.

Floor Plans
Level 1
Level 2
Level 3
Site

33. In the Properties pane, click on the **<None>** button next to View Template.

Identity Data	
View Template	<None>
View Name	Level 2
Dependency	Independent

34. Highlight the **Fire Rating** View Template.

Names:
<None>
Architectural Plan
Fire Rating
Site Plan
Structural Framing Plan

Click **OK**.

35. Activate **Level 3**.

Floor Plans
Level 1
Level 2
Level 3
Site

36. Switch to the View tab on the ribbon.

Select **View Templates→Apply Template Properties to Current View.**

37.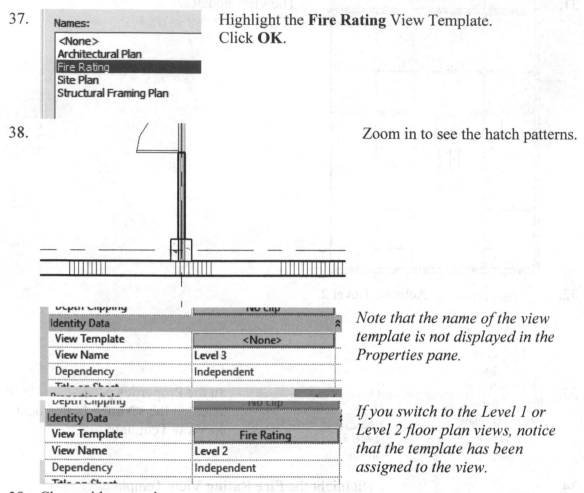

Highlight the **Fire Rating** View Template.
Click **OK**.

38. Zoom in to see the hatch patterns.

Note that the name of the view template is not displayed in the Properties pane.

If you switch to the Level 1 or Level 2 floor plan views, notice that the template has been assigned to the view.

39. Close without saving.

The 1-hr, 2-hr, and 3-hr hatch patterns are custom fill patterns provided inside this exercise file. The *.pat files are included with the exercise files on the publisher's website for your use.

Exercise 4-13

Apply a View Template to a Sheet

Drawing Name: **view templates2.rvt**
Estimated Time to Completion: 10 Minutes

Scope
Use a view template to set all the views on a sheet to the same view scale.
View templates are used to standardize project views. In large offices, different people are working on the same project. By using a view template, everybody's sheets and views will be consistent.

Solution

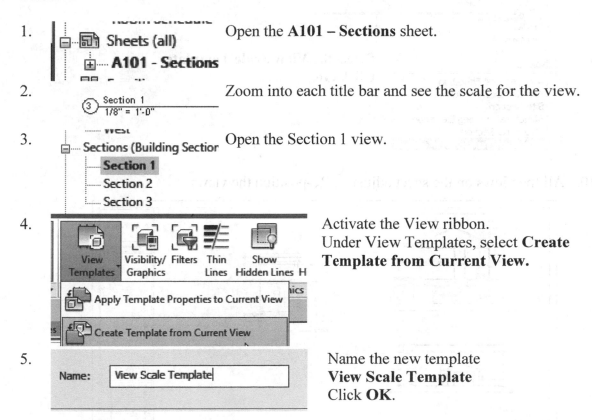

1. Open the **A101 – Sections** sheet.

2. Zoom into each title bar and see the scale for the view.

3. Open the Section 1 view.

4. Activate the View ribbon.
   Under View Templates, select **Create Template from Current View.**

5. Name the new template **View Scale Template**
   Click **OK**.

6. 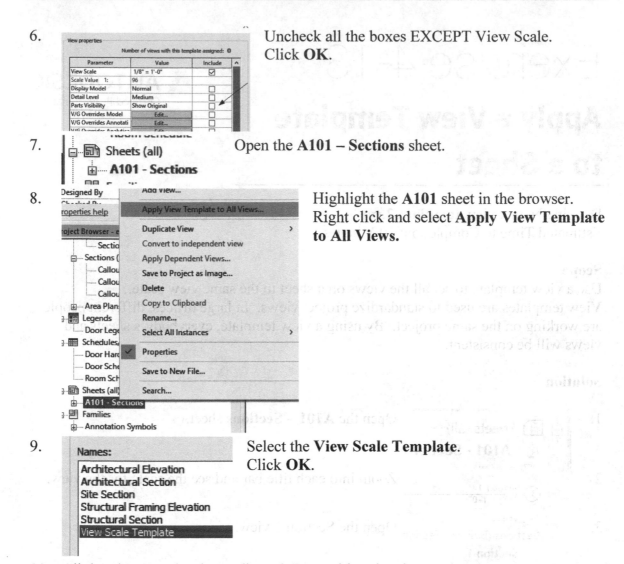   Uncheck all the boxes EXCEPT View Scale.
Click **OK**.

7.   Open the **A101 – Sections** sheet.

8.   Highlight the **A101** sheet in the browser.
Right click and select **Apply View Template to All Views.**

9.   Select the **View Scale Template**.
Click **OK**.

10.   All the views on the sheet adjusted. Re-position the views.

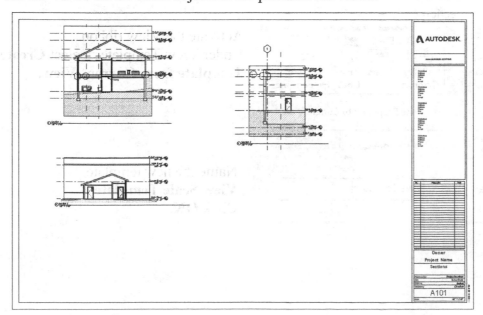

11. 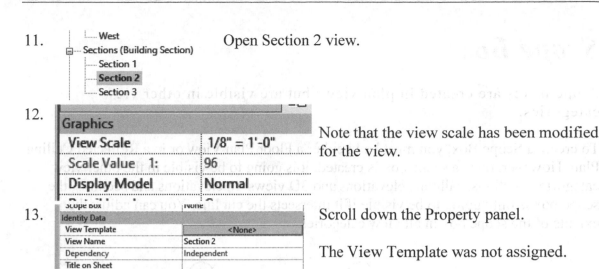Open Section 2 view.

12. Note that the view scale has been modified for the view.

13. Scroll down the Property panel.

The View Template was not assigned.

Click **<None>**.

14. Select the View Scale Template.

Click **OK**.

Scroll up to the View Scale.

Notice that the View Scale is now grayed out.

This is because the View Scale is controlled by the template.

If a parameter in the View Properties is grayed out, check to see if a View Template has been assigned to the view.

15. Close without saving.

Scope Box

Scope boxes are created in plan views but are visible in other view categories.

To create a Scope Box, you must be in either a Floor Plan View or in a Reflected Ceiling Plan. However, once a scope box is created, it is going to be visible in the other view categories: sections, callouts, elevations and 3D views. In elevations and sections, the scope box is only going to be visible if it intersects the cut line. You can adjust the extents of the scope box in all view categories.

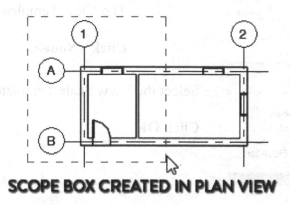

SCOPE BOX CREATED IN PLAN VIEW

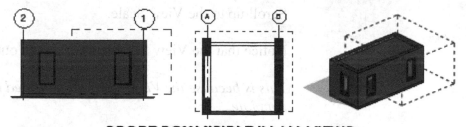

SCOPE BOX VISIBLE IN ALL VIEWS

Consider this office building renovation project. The area affected is in the middle of the building. You want the views to be cropped to fit the red rectangle.

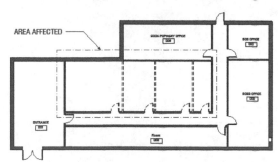

The thing is: you have a lot of views to create. Existing floor plan. Demolished floor plan. New floor plan. Ceilings. Finishes. Layout. All in all, you'll have about 10 views that need the exact same crop region.

An archaic workflow would be to manually adjust the crop region of each view. That would probably work. But what if the project changes and the area affected gets bigger? You have to adjust all the cropped regions again?

That's where the power of scope boxes come into play. Go to the View tab and create a Scope Box. Match it to your intervention area. Give it a name.

Now, apply the scope box to all the views that will be using this cropped region. To save time, select all the views in the project browser by holding the CTRL key.

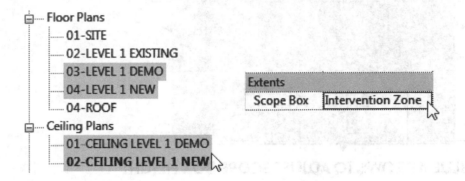

Look at all these views, sharing the exact same crop region. Adjusting the new Scope Box will affect all these views.

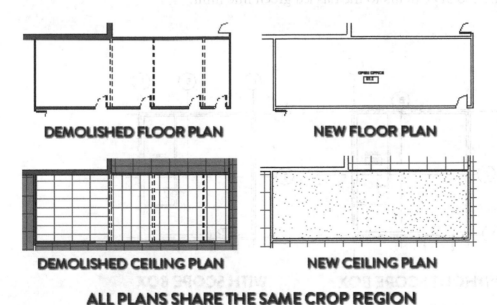

DEMOLISHED FLOOR PLAN

NEW FLOOR PLAN

DEMOLISHED CEILING PLAN

NEW CEILING PLAN

ALL PLANS SHARE THE SAME CROP REGION

Even though Scope Boxes can only be applied in plan view, they are 3D objects. That means they have an assigned height.

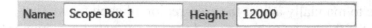

The Height is assigned in the Option bar.

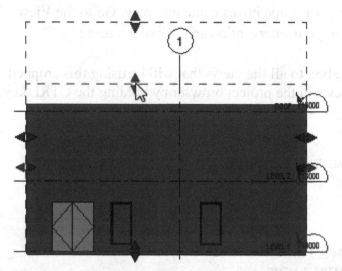

DRAG BLUE ARROWS TO ADJUST SCOPE BOX HEIGHT

Scope boxes can also be used to control the extents and visibility of elements like grids, levels and reference planes. Each of these elements can be assigned to a specific scope box, limiting the 3D extents to the dashed green line limit.

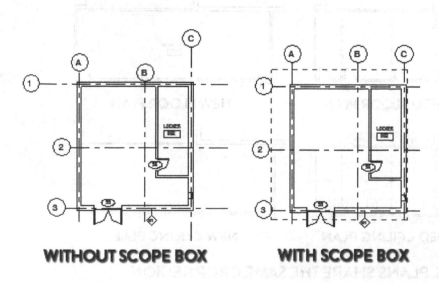

WITHOUT SCOPE BOX   **WITH SCOPE BOX**

Exercise 4-14

Create a Scope Box

Drawing Name: **i_scope_box.rvt**
Estimated Time to Completion: 15 Minutes

Scope
Create and apply a scope box.
Scope boxes can be used to control the visibility of grid lines and levels in views.

Solution

1.
   Activate the **Level 1** floor plan.

2.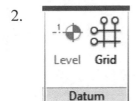
   Activate the **Architecture** tab on the ribbon.
   Select the **Grid** tool from the Datum panel.

3. Offset: 2' 0"
   Set the Offset to **2' 0"** on the Options bar.

4.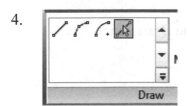
   Select the **Pick Lines** tool from the Draw panel.

5.

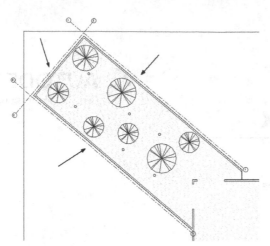

Place three grid lines using the exterior side of the walls to offset.

6.

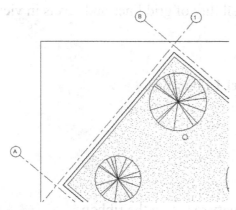

Re-label the grid bubbles so that the two long grid lines are A and B and the short grid line is 1.

7.

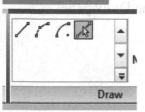

Select the **Grid** tool from the Datum panel.

8. Offset: 180' 0"   Set the Offset to **180' 0"** on the Options bar.

9. Select the **Pick Lines** tool from the Draw panel.

10.

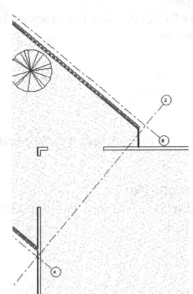

Place the lower grid line by selecting the upper wall and offsetting 180'.
Re-label the grid line **2**.

11.

Activate the **View** tab on the ribbon.
Select the **Scope Box** tool from the Create panel.

12.

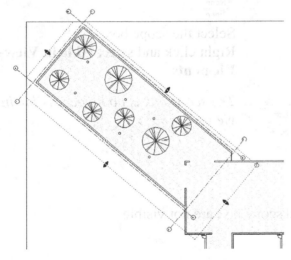

Place the scope box.

Use the Rotate icon on the corner to rotate the scope box into position.

Use the blue grips to control the size of the scope box.

13.

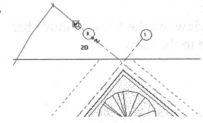

Select the grid line labeled **B**.

14. 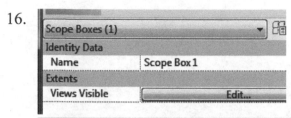 In the Properties pane:
Set the Scope Box to Scope Box 1; the scope box which was just placed.

15. Repeat for the other three grid lines.
You can use the Control key to select more than one grid line and set the Properties of all three at the same time.

16. 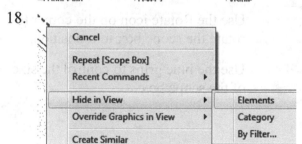 Select the Scope Box.
Select **Edit** next to Views Visible in the Properties pane.

17.

View Type	View Name	Automatic visibility	Override
3D View	{3D}	Visible	None
Ceiling Plan	Level 1	Visible	None
Ceiling Plan	Level 2	Visible	None
Elevation	South Elevation	Invisible	None
Floor Plan	Level 1	Visible	None
Floor Plan	Level 2	Visible	Invisible
Floor Plan	Site	Visible	None
Floor Plan	Level 3	Visible	None

Click on the column headers to change the sort order.

Set the Level 2 Floor Plan Override Invisible.
Click **OK**.

18.

Cancel

Repeat [Scope Box]

Recent Commands ▶

Hide in View ▶ | Elements

Override Graphics in View ▶ | Category

Create Similar | By Filter...

Select the scope box.
Right click and select **Hide in View→ Elements**.

The scope box is no longer visible in the view.

19. Floor Plans
—— Level 1
Level 2
—— Level 3

Activate Level 2.

The grid lines and scope box are not visible.

20. Close the file without saving.

Tip: To make the hide/isolate mode permanent to the view: on the View Control bar, click the glasses icon and then click Apply Hide/Isolate to the view.

Exercise 4-15

Use a Scope Box to Crop Multiple Views

Drawing Name: **i_scope_multiple.rvt**
Estimated Time to Completion: 5 Minutes

Scope
Create and apply a scope box to crop multiple views.

Solution

1.   Views (all)
      Floor Plans
          Level 1
          Level 2
          Level 3
          Level 4
          Level 5
          Level 6
         Activate the **Level 1** floor plan.

2.   Activate the **View** tab on the ribbon.
Select the **Scope Box** tool from the Create panel.

3.   Place the scope box so it encloses the building.

Use the blue grips to control the size of the scope box.

4.   In the Properties panel:

Change the Name of the Scope Box to **Building Model.**

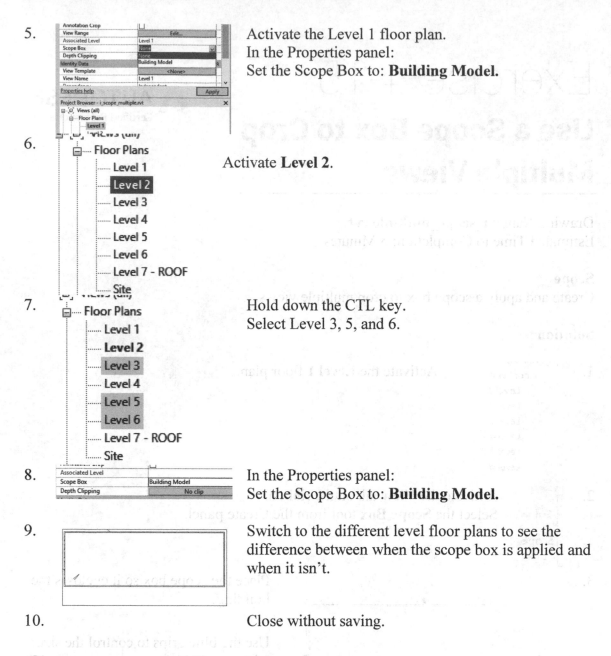

5.   Activate the Level 1 floor plan.
In the Properties panel:
Set the Scope Box to: **Building Model.**

6.   Activate **Level 2**.

7.   Hold down the CTL key.
Select Level 3, 5, and 6.

8.   In the Properties panel:
Set the Scope Box to: **Building Model.**

9.   Switch to the different level floor plans to see the difference between when the scope box is applied and when it isn't.

10.   Close without saving.

Segmented Views

Split a section or elevation view to permit viewing otherwise obscured parts of the view.

This function allows you to vary a section view or an elevation view to show disparate parts of the model without having to create a different view. For example, you may find that landscaping obscures the parts of the model that you would like to see in an elevation view. Splitting the elevation allows you to work around these obstacles.

Exercise 4-16

Segmented Views

Drawing Name: **segmented views.rvt**
Estimated Time to Completion: 10 Minutes

Scope
Modify a section view into segments.

Solution

1.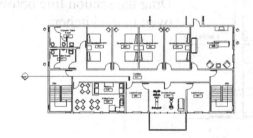

    Activate the **A101- Segmented Elevation** sheet.

    There are two views on the sheet. One is the floor plan and the other is a section view defined in the floor plan.

2.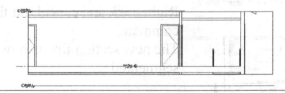

    Right click on the **Main Floor** floor plan view. Select **Activate View**.

    This is the top view on the sheet.

3. Select the section line.

4. 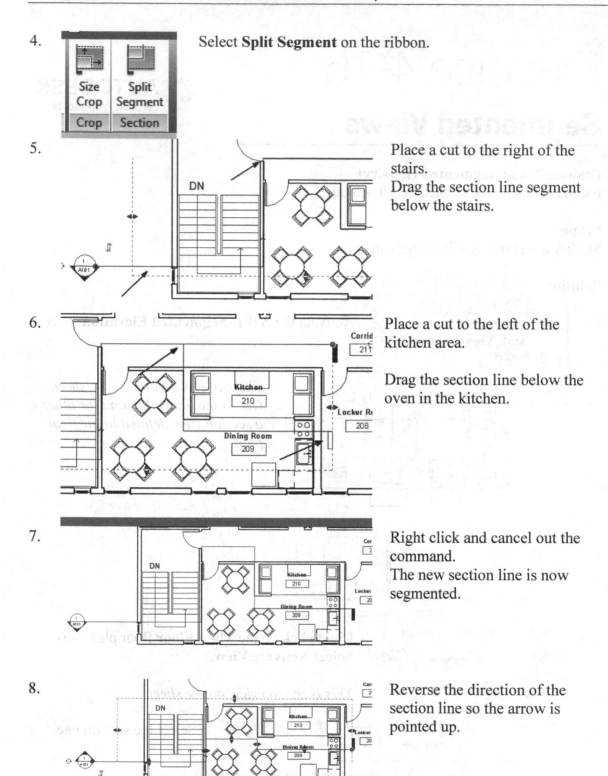 Select **Split Segment** on the ribbon.

5. Place a cut to the right of the stairs.
Drag the section line segment below the stairs.

6. Place a cut to the left of the kitchen area.

Drag the section line below the oven in the kitchen.

7. Right click and cancel out the command.
The new section line is now segmented.

8. Reverse the direction of the section line so the arrow is pointed up.

9. Adjust the segments of the section line so the section lines are as shown.

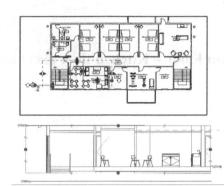

10.

You can see how the new segmented view appears in the lower elevation view.

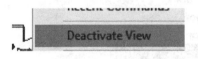

Right click and select **Deactivate View**.

Note how the section view has updated.

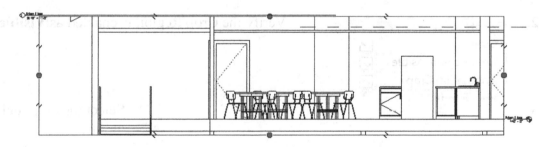

11. Close without saving.

Exercise 4-17

Rotate a Cropped View

Drawing Name: **rotate_view.rvt**
Estimated Time to Completion: 10 Minutes

Scope
To rotate a plan view

Solution

1. Floor Plans
    ⊞ Ground Floor
    ⸺ Lower Roof
    ⊟ Main Floor
    ⸺ Main Floor - Kitchen
    ⸺ Main Roof
    ⸺ Site
    ⸺ T. O. Footing
    ⸺ T. O. Parapet

    Open the **Main Floor – Kitchen** floor plan view.

2. | Extents | |
    | --- | --- |
    | Crop View | ☑ |
    | Crop Region Visible | ☑ |
    | Annotation Crop | ☑ |
    | View Range | |

    Verify the Crop Region is enabled as **Visible**.

3.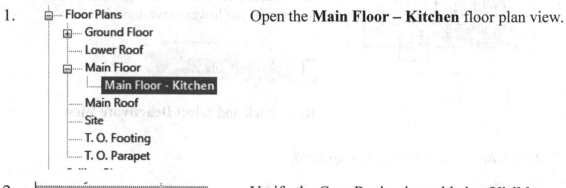

    Select the crop region outline.

4. Select the **Rotate** tool on the Modify panel of the ribbon.

5.

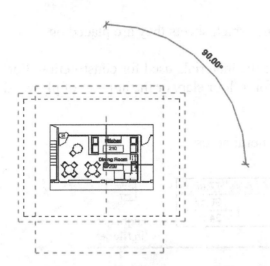

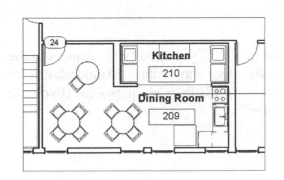

Set the starting point at 0 degrees on the Options bar.

Click the Place button.

Click the center of the view.

6.

Rotate the view 90 degrees.

7.

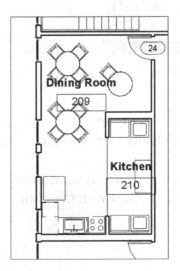

Adjust the crop region's size to show the kitchen and dining room area.

Move the room tags as needed.

Did you notice that the tags rotated with the view?

8.

Close without saving.

Schedules

Schedules are spreadsheets connected to your model. If you modify an element in the model, it will be updated in the schedule and the other way around. Autodesk refers to schedules as "tabulated views", so look for those words in exam questions. If you see those words, the exam is asking about schedules.

Revit offers four different schedule types:

Basic Schedule – used to list and quantify model elements, such as walls, doors, and windows

Sheet & View Lists – used to list views and which sheets they are placed on

Material Takeoff – allows you to calculate the materials used for construction. For example, the amount of concrete needed for a floor slab, or the amount of plywood used in a wall.

Note Block - used to organize plan and general notes

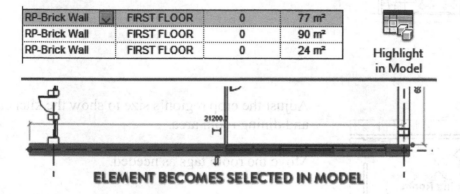

Sometimes, you see an element in a schedule and you need to locate it in the model. Click **Highlight in Model** and a view will open with the element appearing in blue.

When creating a Material Takeoff schedule type, use the Calculated Parameter to calculate values together. For example, multiply the Material Cost with the Material Area to get the total price for each material.

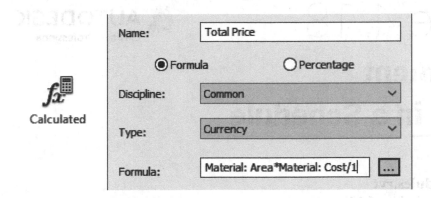

Calculated

Click **Calculated** in the ribbon, select **Currency** and set the **Name**. Then select the two fields in the formula and put * between them. Add a /1 at the end of the formula to fix units, else you will get a warning.

In this example we calculated price, but this tool can be used to calculate anything or create percentages.

Use the Filters tab to exclude specific elements from the schedule. For example, if you want a schedule with walls that are at least 1000mm long, add a «*is greater than or equal*» Length filter. As you see in the resulting schedule, walls below 1000mm are hidden.

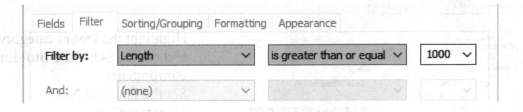

Fields	Filter	Sorting/Grouping	Formatting	Appearance
Filter by:	Length		is greater than or equal	1000
And:	(none)			

<RP-Wall Schedule>				
A	B	C	D	E
Type	Base Constraint	Base Offset	Area	Length
Curtain Wall Ext	FIRST FLOOR	0	5 m²	**1034**
Curtain Wall Ext	FIRST FLOOR	0	6 m²	**1061**
WOOD WALL	FIRST FLOOR	0	10 m²	**7550**

Exercise 4-18
Define Element Properties in a Schedule

Drawing Name: **i_schedules.rvt**
Estimated Time to Completion: 5 Minutes

Scope
Create a door schedule.
Use the sort and group feature to determine how many doors of a specific type are on a level.

Solution

1. 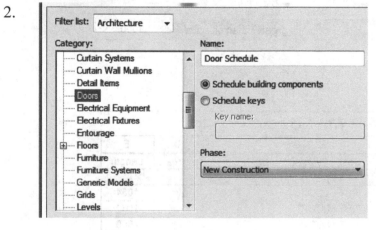 Activate the **View** tab on the ribbon.
   Select **Schedule/Quantities** from the Create panel.

2. Highlight the **Doors** category, and enable **Schedule building components**.
   Set the Phase to **New Construction**.
   Click **OK**.

3.

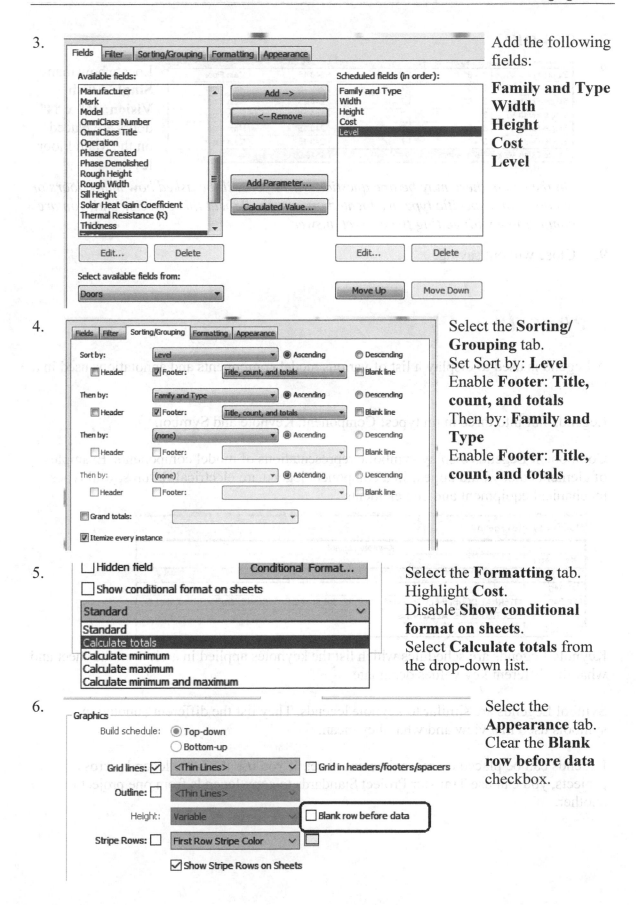

Add the following fields:

Family and Type
Width
Height
Cost
Level

4.

Select the **Sorting/ Grouping** tab.
Set Sort by: **Level**
Enable **Footer**: **Title, count, and totals**
Then by: **Family and Type**
Enable **Footer**: **Title, count, and totals**

5.

Select the **Formatting** tab.
Highlight **Cost**.
Disable **Show conditional format on sheets**.
Select **Calculate totals** from the drop-down list.

6.

Select the **Appearance** tab.
Clear the **Blank row before data** checkbox.

7. Click **OK**.

8.

Single-Flush Vi	3' - 0"	7' - 0"	240.15	Main Floor
Single-Flush Vi	3' - 0"	7' - 0"	240.15	Main Floor
Single-Flush Vi	3' - 0"	7' - 0"	240.15	Main Floor
Single-Flush Vi	3' - 0"	7' - 0"	240.15	Main Floor
Single-Flush Vi	3' - 0"	7' - 0"	240.15	Main Floor
Single-Flush Vi	3' - 0"	7' - 0"	240.15	Main Floor
Single-Flush Vision: 36" x 84": 6			1440.90	

Locate how many **Single-Flush Vision: 36″ x 84″** doors are placed on the Main Floor level.

On the exam, there may be one question where you will be asked how many doors or windows of a specific type are located on a level. Repeat this exercise until you are comfortable with getting the correct answer.

9. Close without saving.

Legends

A Legend is used to display a list of various model components and annotations used in a project.

Legends fall into three main types: Component, Keynote and Symbol.

Component Legends display symbolic representations of model components. Examples of elements that would appear in a component legend are electrical fixtures, wall types, mechanical equipment and site elements.

Modify Keynote Legend	
Keynote Legend1	
Key Value	**Keynote Text**
F23.110	
L20/240	INTERNAL DOUBLE FLUSH TIMBER DOORS, 30MIN FIRE RATED, WITHOUT VISION PANELS
V1	ELECTRICAL INSTALLATIONS
W12.420	

Keynote Legends are schedules which list the keynotes applied in a view on the sheet and what the different key values designate.

Symbol Legends are similar to keynote legends. They list the different annotation symbols used in a view and what they mean.

Legends can be placed on more than one sheet. If you use the same legends across projects, you can use Transfer Project Standards to copy legends from one project to another.

Exercise 4-19

AUTODESK.
Certified Professional

Create a Legend

Drawing Name: **i_Legends.rvt**
Estimated Time to Completion: 30 Minutes

Scope
Create a Legend.
Add to a sheet.
Export Legend.

Solution

1. 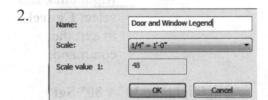   Activate the **View** tab on the ribbon.
Select **Legend** from the Create panel.

2.    Set the Name to **Door and Window Legend**.
Set the Scale to **¼″ = 1′-0″**.
Click **OK**.

3.   The Legend is listed in the browser. An empty view window is opened.

4.    In the browser, locate all the door families.

5.

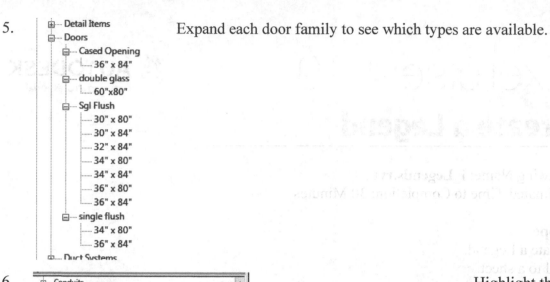

Expand each door family to see which types are available.

6.

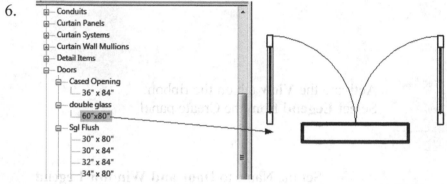

Highlight the **60″ x 80″ Double Glass** door.

Drag and drop it into the Legend view.

Right click and select **Cancel** to exit the command.

7.

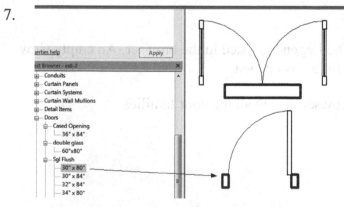

Highlight the **30″ x 80″ Sgl Flush** door.

Drag and drop it into the Legend view.

Right click and select **Cancel** to exit the command.

8.

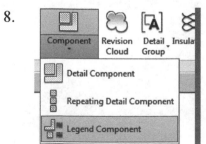

Activate the Annotate tab on the ribbon.

Select the **Legend Component** tool from the Detail panel.

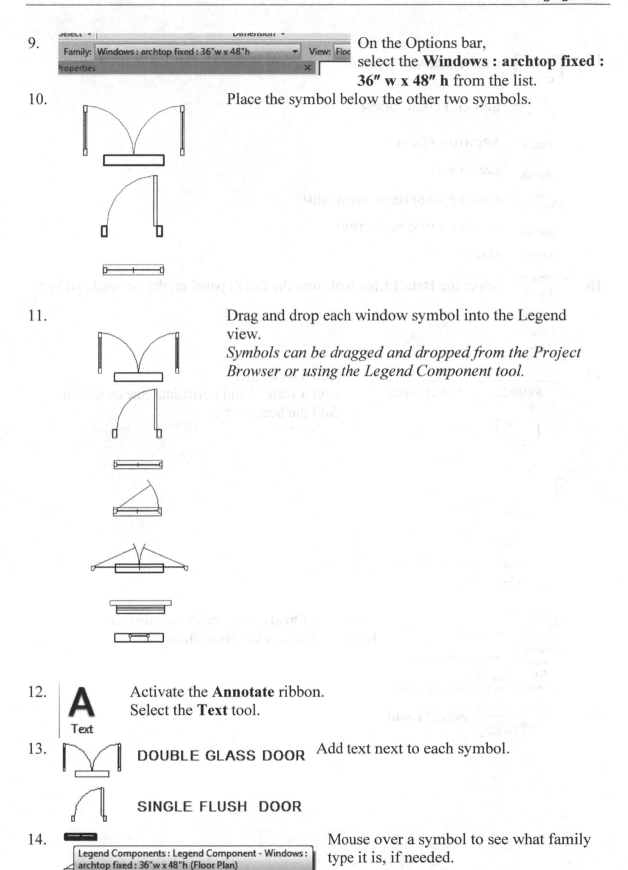

9. On the Options bar, select the **Windows : archtop fixed : 36″ w x 48″ h** from the list.

10. Place the symbol below the other two symbols.

11. Drag and drop each window symbol into the Legend view.
Symbols can be dragged and dropped from the Project Browser or using the Legend Component tool.

12. Activate the **Annotate** ribbon.
Select the **Text** tool.

13. **DOUBLE GLASS DOOR** Add text next to each symbol.

    SINGLE FLUSH DOOR

14. Legend Components : Legend Component - Windows : archtop fixed : 36"w x 48"h (Floor Plan)
Mouse over a symbol to see what family type it is, if needed.

15.

DOUBLE GLASS DOOR

SINGLE FLUSH DOOR

ARCHTOP FIXED

CASEMENT

DOUBLE CASEMENT WITH TRIM

DOUBLE HUNG WITH TRIM

FIXED

Add text as shown.

16.

Detail
Line

Select the **Detail Line** tool from the Detail panel on the Annotate ribbon.

17.

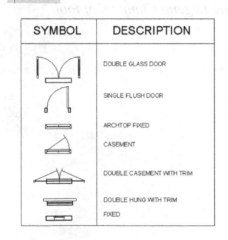

Add a rectangle.
Add a vertical and horizontal line as shown.
Add the header text.

18.

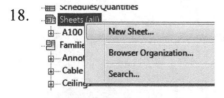

Locate the **Sheets** category in the browser.
Right click and select **New Sheet**.

19.

Load…

Select **Load**.

20. 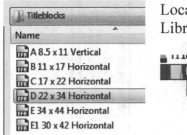 Locate the *Titleblocks* folder under the English - Imperial Library.

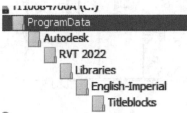

Locate the **D 22 x 34 Horizontal** title block.
Click **Open**.

21. 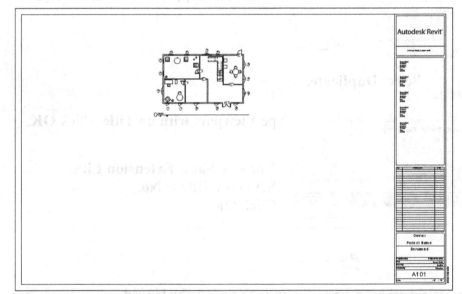 Highlight the **D 22 x 34 Horizontal** title block.
Click **OK**.

22. Drag and drop the Level 1 floor plan onto the sheet.

23. In the Properties pane,
change the View Scale to ¼″ = 1′-0″.

Graphics	
View Scale	1/4" = 1'-0"
Scale Value 1:	48
Display Model	Normal

24. Drag and drop the Door and Window Legend onto the sheet.
Place next to the floor plan view.

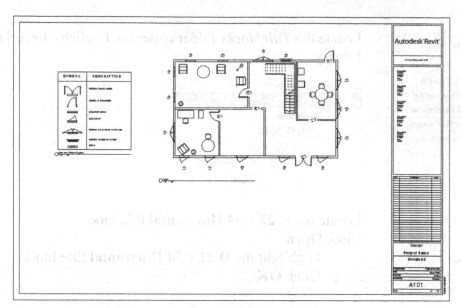

25. Select the legend so it highlights.

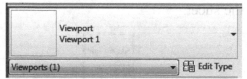

Select **Edit Type** on the Properties pane.

26. Select **Duplicate**.

27. Name: Viewport with no Title

Type **Viewport with no Title**. Click **OK**.

28.

Type Parameters	
Parameter	
Graphics	
Title	view title
Show Title	No
Show Extension Line	☐
Line Weight	1
Color	■ Black
Line Pattern	Solid

Uncheck **Show Extension Line**.
Set Show Title to **No**.
Click **OK**.

29. Zoom in to review the legend.

30. Locate the **Sheets** category in the browser.

New Sheet...

Browser Organization...

Search...

A100 -

A101 -

Families

Annota

Cable Trays

Schedules/Quantities

Sheets (

Right click and select **New Sheet**.

31. Select titleblocks:

D 22 x 34 Horizontal
E1 30 x 42 Horizontal : E1 30x42 Horizontal
None

Highlight the **D 22 x 34 Horizontal** title block.

Click **OK**.

32. Drag and drop the **Level 2** floor plan onto the sheet.

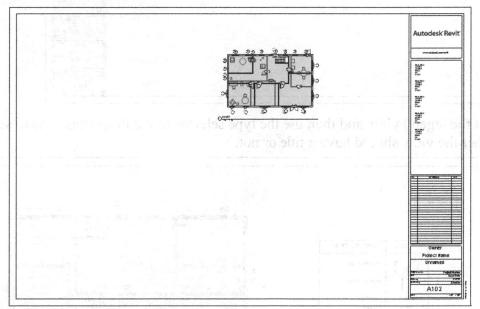

33. Viewports (1)

Graphics

View Scale	1/4" = 1'-0"
Scale Value 1:	48
Display Model	Normal

In the Properties pane,
change the View Scale to **¼″ = 1′-0″**.

34. Drag and drop the Door and Window Legend onto the sheet.
Place next to the view.

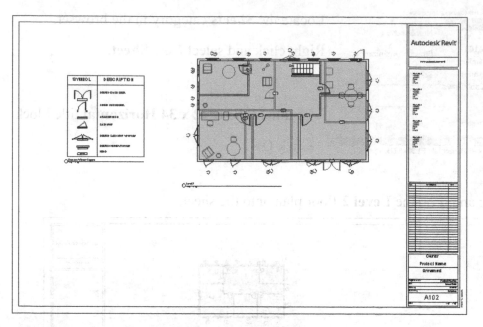

35. Select the legend view and then use the type selector in the Properties pane to set whether the view should have a title or not.

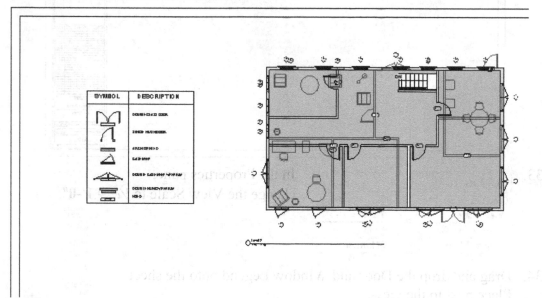

Note that legends can be placed on more than one sheet.

36. Save as *ex4-19.rvt*.

Revisions

Revision tracking is the process of recording changes made to a building model after sheets have been issued.

When working on building projects, changes are often required to meet client or regulatory requirements.

Revisions need to be tracked for future reference. For example, you may want to check the revision history to identify when, why, and by whom a change was made. Revit provides tools that enable you to track revisions and include revision information on sheets in a construction document set.

Use the workflow below for entering and managing revisions. On the exam, you may be asked to organize this workflow or how revision clouds work.

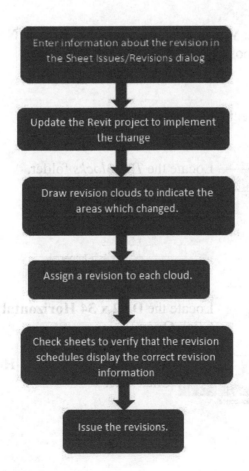

Enter information about the revision in the Sheet Issues/Revisions dialog

Update the Revit project to implement the change

Draw revision clouds to indicate the areas which changed.

Assign a revision to each cloud.

Check sheets to verify that the revision schedules display the correct revision information

Issue the revisions.

Exercise 4-20
Revision Control

Drawing Name: **i_Revisions.rvt**
Estimated Time to Completion: 15 Minutes

Scope
Add a sheet.
Add a view to a sheet.
Setting up Revision Control in a project.

Solution

1.   Activate the **View** tab on the ribbon. Select the **New Sheet** tool on the Sheet Composition panel.

2.   Select the **Load** button.

3.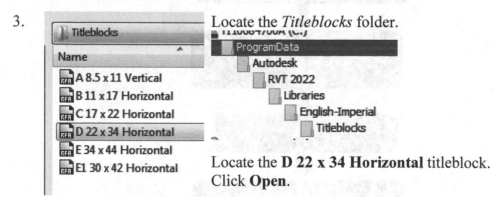

   Locate the *Titleblocks* folder.

   Locate the **D 22 x 34 Horizontal** titleblock.
   Click **Open**.

4.   Highlight the **D 22 x 34 Horizontal** titleblock.
   Click **OK**.

5.

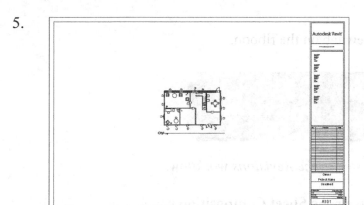

Drag and drop the **Level 1** floor plan onto the sheet.

6.

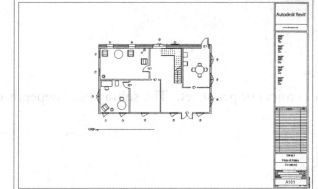

Select the view.
In the Properties pane,
set the View Scale to **1/4″ = 1′-0″**.

7.

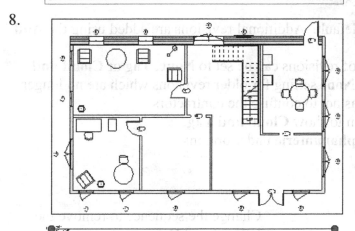

Position the view on the sheet.

8.

To adjust the view title bar, select the view and the grips will become activated.

9. 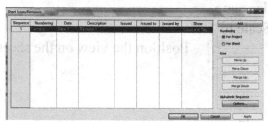 Activate the **View** tab on the ribbon.

> Enter information about the revision in the Sheet Issues/Revisions dialog

This is the first step in the Revisions workflow.

Select **Revisions** on the **Sheet Composition** panel.

10. This dialog manages revision control settings and history.

Numbering can be controlled per project or per sheet. The setting used depends on your company's standards.

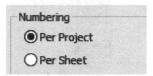

Enable **Per Project**.

One Revision is available by default. Additional revisions are added using the **Add** button.

11. The visibility of revisions can be set to **None**, **Tag** or **Cloud and Tag**. Use the **None** setting for older revisions which are no longer applicable so as not to confuse the contractors.
Set the revision to show **Cloud and Tag**.

12. 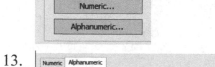 Select **Alphanumeric** under options.

13. Change the sequence to remove the letters **I** and **O**.

Enter sequence values, separated by commas. Each value may be one or more characters. Once all values are used, the sequence will repeat with doubled values.

Sequence: A, B, C, D, E, F, G, H, J, K, L, M, N, P, Q, R, S, T, U, V, W, X, Y, Z

14.

Sequence starting number: [1]

Enter additional characters to display with each value in the sequence.

Prefix: [A]

Suffix: [ ]

Select the **Numeric** tab.
Type **A** for the prefix.
This would allow revisions to be
created using A1, A2, A3…
Click **OK**.

15.

Numbering options

[Numeric…]

[Alphanumeric…]

Enable the **Numeric** option. The dialog will pop up to
show the current settings for Numeric revisions.
Click **OK**.

16. Enter the three revision changes shown.

Use the Add button to add the additional lines.

Sequence	Revision Number	Numbering	Date	Description	Issued	Issued to	Issued by	Show
1	A1	Numeric	08.02	Wall Finish Change	☐	Joe	Sam	Cloud and Tag
2	A2	Numeric	08.04	Window Style Change	☐	Joe	Sam	Cloud and Tag
3	A3	Numeric	08.13	Door Hardware Change	☐	Joe	Sam	Cloud and Tag

17. Click **OK** to close the dialog.
You can delete revisions if you make a mistake. Just highlight the row and click
Delete.

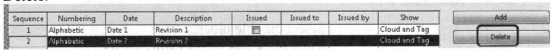

Sequence	Numbering	Date	Description	Issued	Issued to	Issued by	Show		
1	Alphabetic	Date 1	Revision 1	☐			Cloud and Tag	Add	
2	Alphabetic	Date 2	Revision 2				Cloud and Tag	Delete	

18. Save the project as *ex4-20.rvt*.

Exercise 4-21
Modify a Revision Schedule

Drawing Name: **revision_schedule.rvt**
Estimated Time to Completion: 20 Minutes

Scope
Modify a revision schedule in a title block.

Solution

1. Zoom in to the **Revision Block** area on the sheet.
Note that the title block includes a revision schedule
by default.

2. 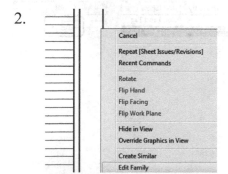 Select the title block.
Right click and select **Edit Family**.

3. 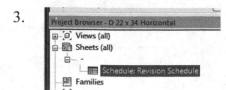 Select the **Revision Schedule** in the Project Browser.

4. 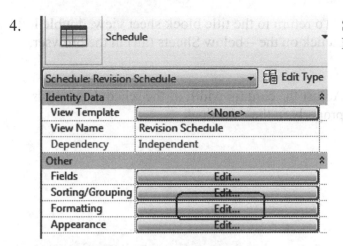   Select **Edit** next to Formatting in the Properties pane.

5. 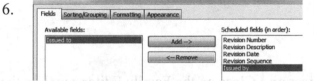   Change the First Column Header to **Rev**.

6.   Select the Fields tab.
      Add the **Issued By** field.

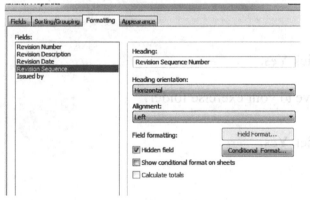

Do **NOT** remove the Revision Sequence field. This is a hidden field.

Note that the Hidden field control is located on the Formatting tab. This is a possible question on the certification exam.

7.   Order the fields as shown.
      Click **OK**.

Scheduled fields (in order):

Revision Number
Revision Description
Revision Date
Issued by
Revision Sequence

8.

A	B	C	D
Rev	Description	Date	Issued by

The Revision Schedule updates.

9.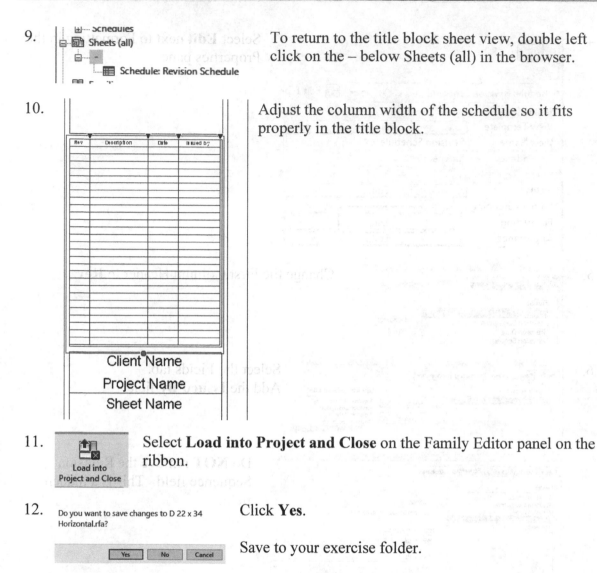

To return to the title block sheet view, double left click on the – below Sheets (all) in the browser.

10.

Adjust the column width of the schedule so it fits properly in the title block.

11. Select **Load into Project and Close** on the Family Editor panel on the ribbon.

12. Do you want to save changes to D 22 x 34 Horizontal.rfa?

Click **Yes**.

Yes   No   Cancel

Save to your exercise folder.

13. Replace Existing File   ×

The file D 22 x 34 Horizontal.rfa already exists. Do you want to replace the existing file?

Yes   No   Cancel

Click **Yes**.

14. You are trying to load the family D 22 x 34 Horizontal, which already exists in this project. What do you want to do?

→ Overwrite the existing version

→ Overwrite the existing version and its parameter values

Click on **Overwrite the existing version and its parameter values**.

15.

Rev	Description	Date	Issued by

Owner

Note the title block updates with the new revision schedule format.

Save as *ex4-21.rvt*.

Step Two of the workflow

Update the Revit project to implement the change

Description
Wall Finish Change
Window Style Change
Door Hardware Change

We have three changes to make in our project.

Exercise 4-22

Update Project

Drawing Name: **update_project.rvt**
Estimated Time to Completion: 5 Minutes

Scope
Make changes to a Revit project using Phases.
Change a wall finish.
Change a window style.
Change a door style.

Solution

1.
    ⊟···· 3D Views
        ····· **Interior - Level 1**
        ····· Southeast
        ····· {3D}

Activate the **Interior - Level 1** 3D view.

2.
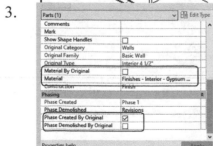

If you hover over an interior wall, you will see that the wall has been divided into parts.

This allows you to replace the finish on the walls with a different finish.

Select the wall part.

3.

Parts (1)	Edit Type
Comments	
Mark	
Show Shape Handles	☐
Original Category	Walls
Original Family	Basic Wall
Original Type	Interior 4 1/2"
Material By Original	☐
Material	Finishes - Interior - Gypsum ...
Construction	Finish
Phasing	
Phase Created	Phase 1
Phase Demolished	Revisions
Phase Created By Original	☑
Phase Demolished By Original	☐

Uncheck Material By Original.

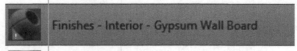

Change the Material to **Finishes – Interior – Gypsum Wall Board.**
Click **OK.**
Uncheck **Phase Demolished By Original.**

4.
    ⊟····· Floor Plans
        ······ **Level 1**
        ······ Level 1 –Revisions
        ······ Level 2
        ······ Roof

Activate the **Level 1** floor plan.

5.

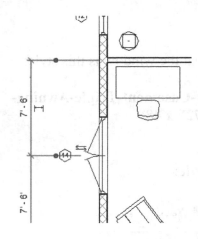

Select Window 14 located on the west wall.

Mark	6
Phasing	
Phase Created	Phase 1
Phase Demolished	Revisions

In the Properties panel:
Verify that it was demolished in the Revisions phase.

Click **ESC** to release the selection.

6.

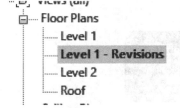

Select Door 3.

Mark	6
Phasing	
Phase Created	Phase 1
Phase Demolished	Revisions

In the Properties panel:
Verify that it was demolished in the Revisions phase.

Click **ESC** to release the selection.

7.

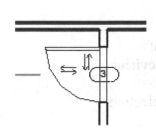

- Views (all)
  - Floor Plans
    - Level 1
    - **Level 1 - Revisions**
    - Level 2
    - Roof

Activate the **Level 1 – Revisions** floor plan.

8.

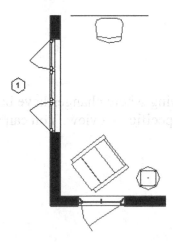

Select window 1 on the west wall.

Referencing Detail	
Phasing	
Phase Filter	Show Previous + New
Phase	Revisions

Notice that the window was created in the Revisions phase.

9.

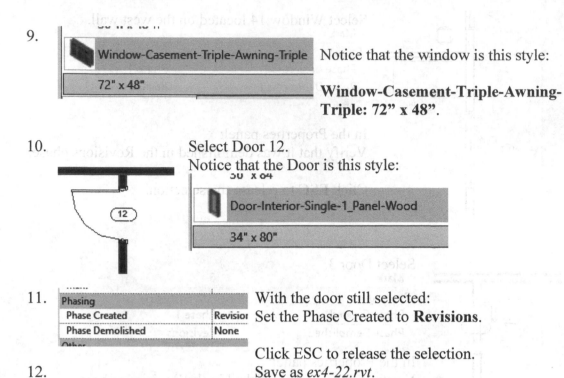

Notice that the window is this style:

Window-Casement-Triple-Awning-Triple: 72" x 48".

10. Select Door 12.
Notice that the Door is this style:

11. With the door still selected:
Set the Phase Created to **Revisions**.

Click ESC to release the selection.

12. Save as *ex4-22.rvt*.

Step Three of the workflow

Draw revision clouds to indicate the areas which changed.

Revision Clouds

Revision clouds are used to designate areas in the drawing where changes have been made. Revision clouds are annotation objects and are specific to a view. You can assign the same revision to multiple clouds.

Exercise 4-23

Add Revision Clouds

Drawing Name: **revision clouds.rvt**
Estimated Time to Completion: 10 Minutes

Scope
Add revision clouds to a view.

Solution

1.   Activate the Unnamed sheet with the **Level 1** floor plan.

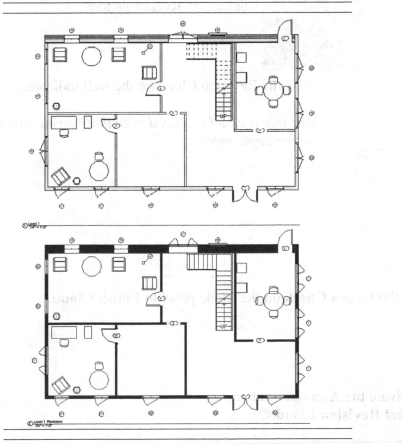

The top view shows the existing floor plan prior to the revisions.
The bottom view shows the revisions that were made.
We will be adding revision clouds to the bottom view.

2. Activate the **Annotate** ribbon.
Select **Revision Cloud** from the Detail panel.

3. 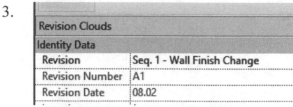 On the Properties pane,
select the Revision that is tied to the
revision cloud – **Wall Finish Change**.

Revision Clouds	
Identity Data	
Revision	Seq. 1 - Wall Finish Change
Revision Number	A1
Revision Date	08.02

4. Under Comments, enter **Finish Changed from plaster to gypsum wallboard**.

Revision Clouds		Edit Type
Identity Data		
Revision	Seq. 1 - Wall Finish Change	
Revision Number	A1	
Revision Date	08.02	
Issued to	Joe	
Issued by	Sam	
Mark		
Comments	Finish changed from plaster to gypsum wallboard	

5. 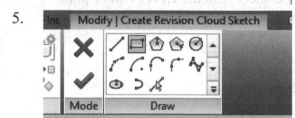 You can use any of the available Draw
tools to create your revision cloud.
Select the **Rectangle** tool.

6. 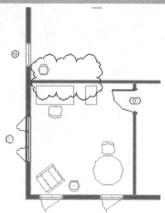 Draw the Revision Cloud on the wall indicated.

Note that a cloud is placed even though you selected the rectangle tool.

7.  Select the **Green Check** on the Mode panel to **Finish Cloud**.

8. Activate the **Annotate** ribbon.
Select **Revision Cloud**.

9. 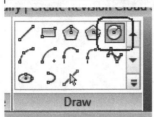 Select the Revision that is tied to the revision cloud – **Window Style Change**.

Under Comments, enter **Use Assy Code 301**.

10. 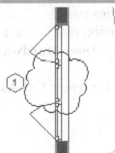 Select the **Circle** tool.

11. Draw the Revision Cloud on the window indicated.

Note that a cloud is placed even though you selected the circle tool.

12. 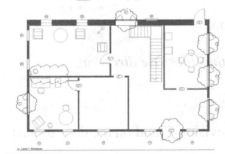 Add revision clouds to the other windows that were changed.

Look for windows tagged with the number 1.

13. Select the **Green Check** on the Mode panel to **Finish Cloud**.

14.

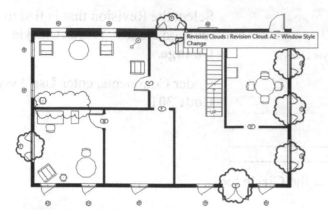

If you hover over the revision cloud, you will see a tooltip to indicate what the revision is.

Notice all the revision clouds highlight.

15.

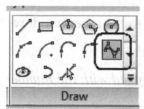

Activate the **Annotate** ribbon.
Select **Revision Cloud** from the Detail panel.

16.

Revision Clouds	
Identity Data	
Revision	Seq. 3 - Door Hardware Change
Revision Number	A3
Revision Date	08.13
Issued to	Joe
Issued by	Sam
Mark	
Comments	Use Assy Code 405.

On the Properties pane,
select the Revision that is tied to the revision cloud – **Door Hardware Change**.
Under Comments, enter **Use Assy Code 405**.

17.

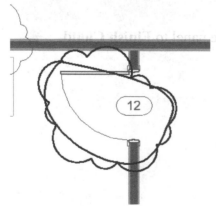

Select the **Spline** tool.

Draw the cloud in the clockwise direction.

18.

Draw the Revision Cloud on the door indicated.

19. **Error - cannot be ignored**

Self intersecting splines are not allowed in sketches.

Show

You might see this error if the spline has overlapping segments. If you get this error, delete the spline and try again. You do not have to close the spline in order to create the revision cloud.

20. Select the **Green Check** on the Mode panel to **Finish Cloud**.

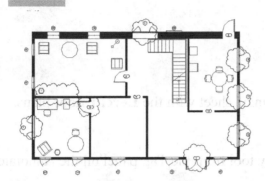

The revision clouds have been placed in the view.

21.

Rev	Description	Date	Issued by
A1	Wall Finish Change	08.02	Sam
A2	Window Style Change	08.04	Sam
A3	Door Hardware Change	08.13	Sam

Zoom into the title block and note that the revision block has updated with the revisions that have been issued.

22. Save as *ex4-23.rvt*.

Step Four of the workflow

Assign a revision to each cloud.

Exercise 4-24

AUTODESK.
Certified Professional

Tag Revision Clouds

Drawing Name: **tag clouds.rvt**
Estimated Time to Completion: 5 Minutes

Scope
Tag revision clouds in a view.

Solution

1. Activate the Unnamed sheet with the **Level 1** floor plans.

   Sheets (all)
   A100 - Cover Sheet
   A101 - Unnamed
   Families

2. Select the **Tag by Category** tool from the Tag panel on the Annotate tab on the ribbon.

   Tag by Category

3.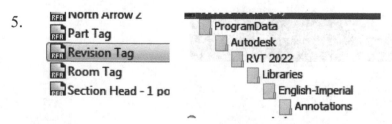

   Revision Clouds : Revision Cloud: A2 - Window Style Change

   Pick one of the revision clouds to identify the category to be tagged.

4. There is no tag loaded for Revision Clouds. Do you want to load one now?

   If you see this message, click **Yes**.

5. North Arrow 2
   Part Tag
   Revision Tag
   Room Tag
   Section Head - 1 po

   ProgramData
   Autodesk
   RVT 2022
   Libraries
   English-Imperial
   Annotations

   Select the *Annotations* folder.
   Locate the **Revision Tag**.
   Click **Open**.

6. Horizontal | Tags... | ☑ Leader | Attached End | 1/2"

   Enable **Leader** on the Option bar.

7. Select the revision cloud to add the tag. The tag is placed.

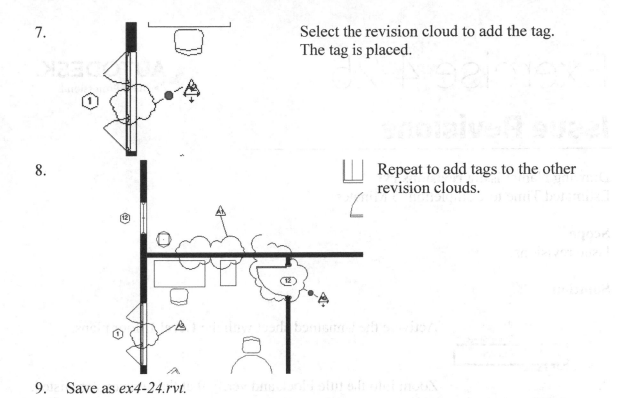

8. Repeat to add tags to the other revision clouds.

9. Save as *ex4-24.rvt*.

Step Five & Six of the workflow

Check sheets to verify that the revision schedules display the correct revision information

Issue the revisions.

Exercise 4-25

Issue Revisions

Drawing Name: **issue revisions.rvt**
Estimated Time to Completion: 5 Minutes

Scope
Issue revisions.

Solution

1. Activate the Unnamed sheet with the **Level 1** floor plans.

   Sheets (all)
   A100 - Cover Sheet
   A101 - Unnamed
   Families

2. Zoom into the title block and verify that the revisions are listed.

3. Switch to the View tab on the ribbon.

   Revisions

   Select **Revisions**.

4. Place a check on the Issued button for each revision.

   Note that they are now grayed out.

   Click **OK**.

5. Select one of the clouds.

   Notice that the revision comment now shows (Issued).

   Revision Clouds : Revision Cloud: A1 - Wall Finish Change (Issued)

6. In the Properties pane, the Revision now displays *(Issued)*.

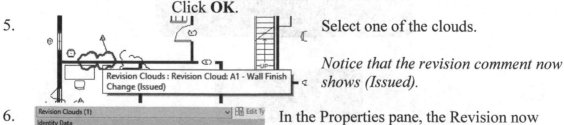

7. Save as *ex4-25.rvt*.

Sheet Lists

One of the types of schedules available in Revit is sheet lists. A sheet list schedule is built the same way as a component model schedule, but all the relevant parameters are used by sheets. Creating a sheet list is useful for managing your construction documentation and organizing any documentation packages you sent out as it operates as a table of contents.

If you are required to submit a list of all drawings in a submittal package, you can use a Sheet List schedule.

As with most schedules that appear on construction documents, it is good practice to have two different versions of each schedule – one that has all the parameters necessary for tracking the sheets and any revisions and another that is used for the actual documentation package.

Exercise 4-26

Sheet Lists

Drawing Name: *sheet_lists.rvt*
Estimated Time: 5 minutes

This exercise reinforces the following skills:
- ❑ Schedules
- ❑ Sheets

1. 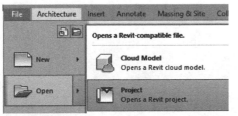 Go to **File**.

    Select **Open→Project**.
    Open *sheet_lists.rvt*.

2.   Activate the **View** tab on the ribbon.

3.

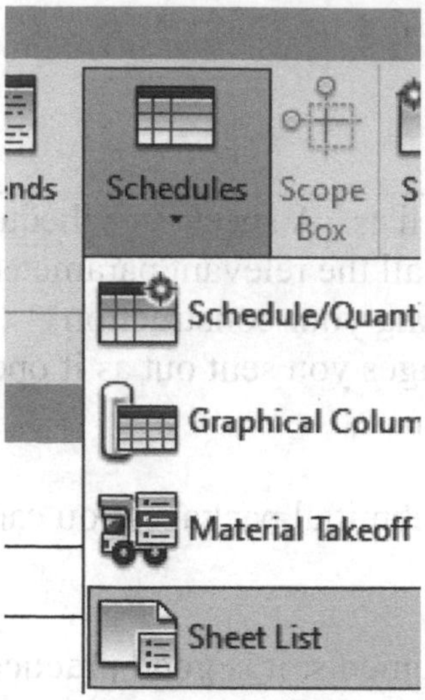

Go to **Schedules→Sheet List.**

4.

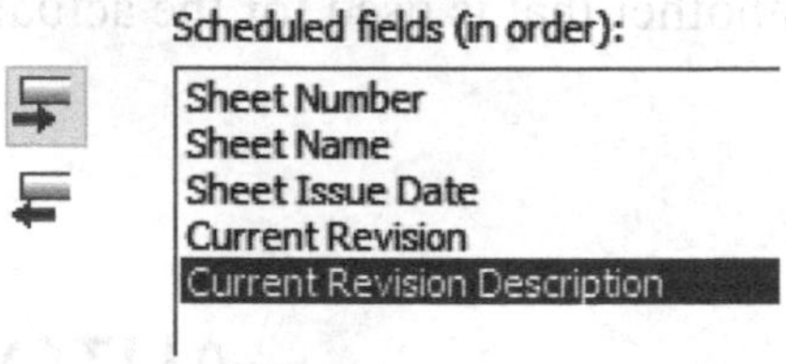

Select the following fields in order:

- Sheet Number
- Sheet Name
- Sheet Issue Date
- Current Revision
- Current Revision Description

Click **OK**.

5.

A	B	C	D	E
Sheet Number	Sheet Name	Sheet Issue Date	Current Revision	Current Revision D
A102	Plans	07/24/12		
A104	Elev./Sec./Det.	07/30/12		
A103	Elevations/Section	11/15/12		
A105	Elev./ Stair Section	11/15/12		
A101	Site Plan	11/16/12		
A001	Title Sheet	11/16/12		

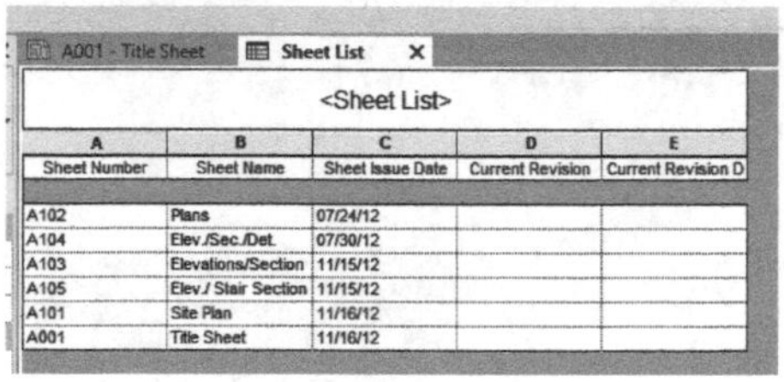

The schedule view opens.
Save as *ex4-26.rvt.*

Exercise 4-27

Aligning Views Between Sheets

Drawing Name: **aligning_views.rvt**
Estimated Time to Completion: 30 Minutes

Scope
Using Grid Guides

Solution

1.  Activate the sheet with **Level 1** floor plan.

2. In the Properties pane:
   Scroll down and note that Grid Guide is set to
   <None>.

3. Activate the **View** ribbon.
   Select **Guide Grid** on the Sheet Composition panel.

   s Guide
   Grid

4.  Click **OK**.

5.

Dimensions	
Guide Spacing	2"
Identity Data	
Name	Guide Grid 1

Select the Guide Grid.
Select by left clicking on an edge.
In the Properties pane,
set the Guide Spacing to **2"**.
Click **Apply**.

6.

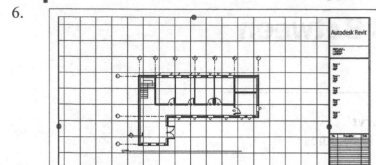

The grid updates.
Use the blue grips to adjust
the grid so it lies entirely
inside the title block.

7.

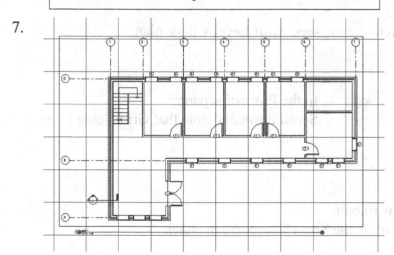

Select the viewport.

Select the **Move** tool from
the Modify Panel.

8.

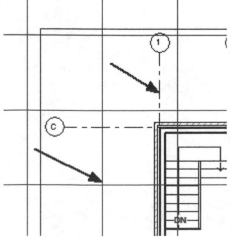

Select on the grid line 1 in the view.
Then select the guide grid.

9. 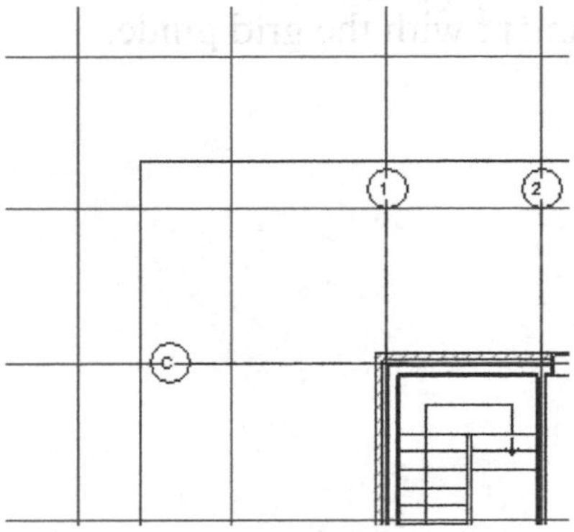   The view snaps to align with the grid.
     Repeat to shift grid line C into alignment with the grid guide.

10. 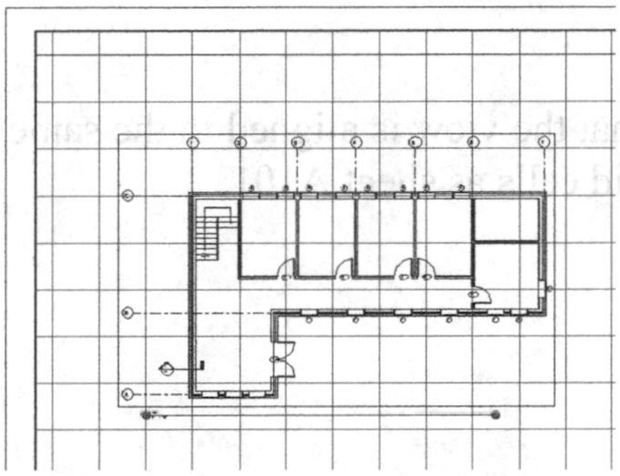 Zoom out and note which grid the view is aligned to.

11. 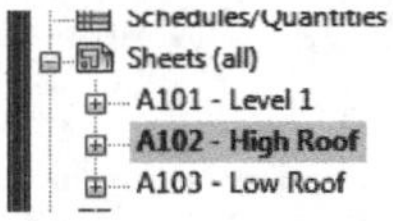 Schedules/Quantities
     Sheets (all)
       A101 - Level 1
       A102 - High Roof
       A103 - Low Roof

     Activate the sheet named **High Roof**.

12.  | Other | |
     |---|---|
     | File Path | |
     | Drawn By | Author |
     | Guide Grid | Guide Grid 1 |

     In the Properties pane,
     set the Guide Grid to **Guide Grid 1**.

13.  Select the viewport.

     Select the **Move** tool from the Modify Panel.

14. 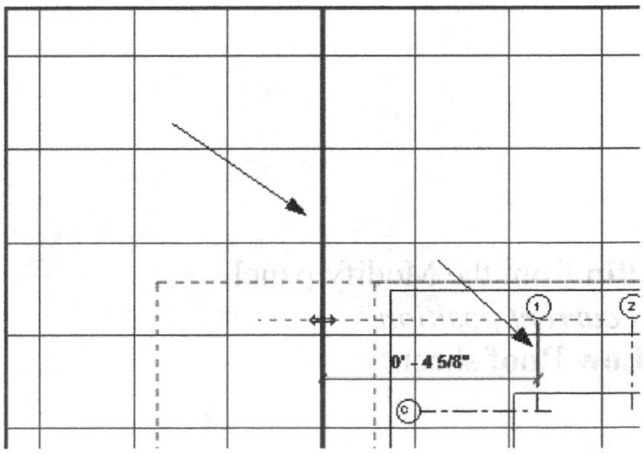 Select on the grid line 1 in the view.
     Then select the guide grid.

     0' - 4 5/8"

15.

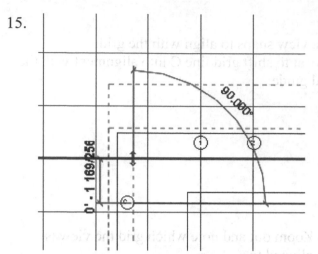

Repeat to shift grid line C into alignment with the grid guide.

16.

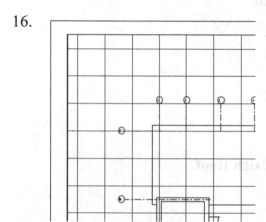

Verify that the view is aligned to the same guide grid cells as sheet A101.

17.

Type **VV**. Select the **Annotation Categories** tab. Disable visibility of the **Guide Grid**.
Click **OK**.

18. Select the viewport. Select **Pin** from the Modify panel.
This will lock the view to its current position.

19. Activate the **Low Roof** sheet.

Sheets (all)
  A101 - Level 1
  A102 - High Roof
  A103 - Low Roof
Families

20. Select **Guide Grid** from the View ribbon.

Guide
Grid

21. *Note you can select the existing guide grid.*

Assign Guide Grid ☒

⦿ Choose existing:

Guide Grid 1

○ Create new:

Name: Guide Grid 2

Click **OK**.

22. Note in the Properties pane the Guide Grid 1 is listed.

Other	
File Path	
Drawn By	Author
Guide Grid	Guide Grid 1

23. Rotation on Sheet: 90° Clockwise

Select the view. On the Options bar, set the Rotation on Sheet to **90° Clockwise**.

24. The view rotates.

25. Close the file without saving.

Certified User Practice Exam

1. Straight grid lines are visible in the following view types:

   A. ELEVATION
   B. PLAN
   C. 3D
   D. SECTION
   E. DETAIL

2. Which two shortcut keys launch the Visibility/Graphics dialog?

   A. VG
   B. VV
   C. VE
   D. VW
   E. F5

3. **True or False**: Objects hidden using the Temporary Hide/Isolate tool are not visible, but they are still printed.

4. Select the THREE options for Detail Level for a view:

   A. COARSE
   B. SHADED
   C. FINE
   D. HIDDEN
   E. WIREFRAME
   F. MEDIUM

5. **True or False**: If you delete a view, the annotations placed in the view are also deleted.

6. **True or False**: A camera cannot be placed in an elevation view.

7. To change the graphic appearance of your model from Hidden Line to Realistic, you:

   A. Modify the Rendering Settings in the View Control Bar
   B. Edit Visibility/Graphics Overrides
   C. Change graphic display options in View Properties
   D. Click Visual Styles on the View Control Bar

8. A story level is the color:

    A. Yellow
    B. Blue
    C. Black
    D. Green

9. If you create a level using the COPY or ARRAY tool, what type of level is created?

    A. Story
    B. Non-Story or Reference
    C. Elevation
    D. Plan

10. This type of level does not have a PLAN view associated to it:

    A. Story
    B. Non-Story or Reference
    C. Elevation
    D. Plan

11. In which view type can you place a level?

    A. PLAN
    B. LEGEND
    C. 3D
    D. CONSTRUCTION
    E. ELEVATION

12. To display or open a view: (Select all valid answers)

    A. Double click the view name in the project browser
    B. Double click the elevation arrowhead, the section head or the callout head
    C. Right click the elevation arrowhead, the section head or the callout head, and select Go to View
    D. Select the Surf tool from the ribbon

13. Select the icon used to lock a 3D view

Certified Professional Practice Exam

1. Scope boxes control the visibility of:

    A. Elements
    B. Plumbing Fixtures
    C. Object Styles
    D. Grid lines and levels

2. If you rotate a cropped view: (Select the answer which is TRUE)

    A. Any tags will automatically rotate with the view.
    B. The crop region will automatically resize.
    C. The Project North will change.
    D. Any hidden elements will become visible.

3. Depth Cueing applies to all the elements listed EXCEPT: (Select two answers)

    A. Model elements, like walls, doors, and floors.
    B. Shadows and Sketchy Lines
    C. Annotations, like dimensions and tags
    D. Linework and line weight

4. A _____ is a collection of view properties, such as view scale, discipline, detail level and visibility settings.

    A. View template
    B. Graphics display
    C. Crop Region
    D. Sheet

5. _____ provide(s) a way to override the graphic display and control of the visibility of elements that share common properties in a view.

    A. Hide/Isolate
    B. Hide Category
    C. Hide Element
    D. Filters

6. Select the Type properties for a Level. (Select all correct answers)

    A. Elevation Base
    B. Line Weight
    C. Elevation
    D. Name

7. Select the ONE item you cannot create a view type:

    A. Floor Plans
    B. Ceiling Plans
    C. Area Plans
    D. 3D Views

8. A _____ displays the view number and sheet number (if the view is included on a sheet) in the corresponding view.

    A. Title bar
    B. View Name
    C. Sheet
    D. View Reference

Floor Plan: Level 1	
Graphics	
View Scale	1/8" = 1'-0"
Scale Value 1:	96
Display Model	Normal
Detail Level	Coarse
Parts Visibility	Show Original
Detail Number	1
Rotation on Sheet	None

9. The View Scale, Detail Level, and Display is grayed out for the floor plan view. The reason is:

    A. The view is locked
    B. The view is pinned
    C. A view template has been assigned to the view
    D. The view range is applied.

10. A scope box is used to: (select more than one)

    A. Section a 3D view
    B. Crop multiple views
    C. Control the visibility of grid and/or levels in different views
    D. Control the depth of a 3D camera

11. T/F
Once a revision is issued, it cannot be modified.

12. T/F
A revision can only be applied to one revision cloud.

13.　T/F
　　　Legends can be placed on more than one sheet.

14.　Select the icon used to reveal hidden elements.

15.　Select the icon used to reveal hidden elements.

16.　T/F
　　　You place an annotation in a dependent view. It will not be visible in the parent view.

17.　The tool to align views between sheets is called:

　　　A.　Align
　　　B.　Guide Grid
　　　C.　Align Grid
　　　D.　Pin

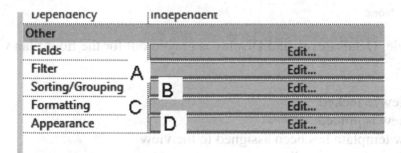

18.
　　　To only see the doors placed on Level 1, which tab should be used?

Documentation

This lesson addresses the following certification exam questions:

- Multi-Segmented Dimensions
- Dimensional Constraints
- Global Parameters
- Project Parameters
- Shared Parameters
- Filled Regions
- Detail Components
- Repeating Details
- Adding Tags
- Color Schemes
- Color Scheme Legends
- Design Options
- Browser Organization for Sheets
- Phases

Dimensions are system families. They are in the Annotation category. They have type and instance properties.

Revit has three types of dimensions: listening, temporary and permanent.

A listening dimension is the dimension that is displayed as you are drawing, modifying, or moving an element.

A temporary dimension is displayed when an element is placed or selected. In order to modify a temporary dimension, you must select the element.

A permanent dimension is placed using the Dimension tool. In order to modify a permanent dimension, you must move the element to a new position. The permanent dimension will automatically update. To reposition an element, you can modify the temporary dimension or move the element using listening dimensions.

When you enter dimension values using feet and inches, you do not have to enter the units. You can separate the feet and inches values with a space and Revit will fill in the units. If a single unit is entered, for example '10', Revit assumes that value is 10 feet, not 10 inches.

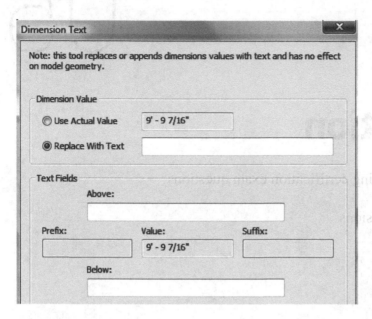

To override a dimension value, select the permanent dimension and enable Replace with Text and put in the desired text.

You can also add additional notes using the Text Fields for Above, Prefix, Suffix, or Below.

Revit has the capability of including symbols in textual notes. It is highly likely to come up as an exam question on the user or professional certification exams, so be prepared to demonstrate how to add a symbol to a note.

Exercise 5-1

Create Dimensions

Drawing Name: **i_dimensions.rvt**
Estimated Time to Completion: 20 Minutes

Scope
Placing dimensions

Solution

1.
   - Floor Plans
     - Ground Floor
     - **Ground Floor Admin Wing**
     - Lower Roof
     - Main Floor
     - Main Floor Admin Wing

   Activate the **Ground Floor Admin Wing** floor plan.

2. Activate the Modify tab on tab on ribbon.
   Select the **Match Properties** tool from the Clipboard panel.

3.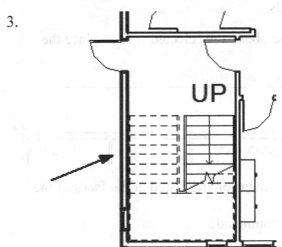

   Select the wall indicated as the source object.

4. Select the wall indicated as the target object.

5. Note that the curtain wall changed to an Exterior-Siding wall.

   Cancel out of the Match Properties command.

6. Activate the Annotate tab on tab on ribbon.
   Select the **Aligned Dimension** tool.

   Aligned I

7. On the Options bar, set the dimensions to select the **Wall faces**.

   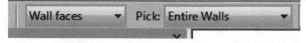

8. Set the dimensions to **Pick Entire Walls**.

9. Pick the wall indicated. Move the mouse above the selected wall to place the dimension.

   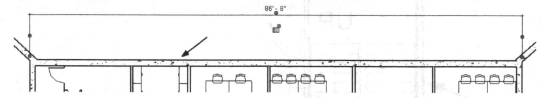

10. Note the entire wall is selected and the dimension is located at the faces of the walls.
    Cancel or escape to end the Dimension command.

11. Select the dimension.
    Note that there are several grips available. The grips are the small blue bubbles.

   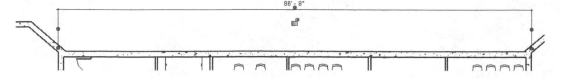

12. 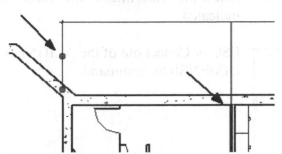 Select the middle grip indicated on the left.
Move the witness line to the wall face indicated by the arrow in the center of the image.

Note that the witness line automatically snaps to the wall face.

13. 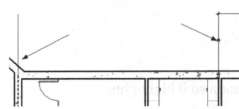 With the dimension selected,
on the Options bar, change the Prefer to **Wall centerlines**.

14. 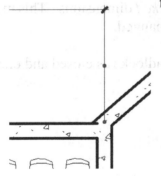 Select the dimension and activate the grip indicated.
Drag the witness line to the center of the left wall.

15. Left click once on the grip on the right side of the dimension to shift the witness line to the wall centerline.

16. *Note how the dimension value updates.*

17.  Select the **Aligned** tool.

18. Set **Pick to Individual References** on the options bar.

19. Select the centerlines of the walls indicated.

ESC or Cancel out of the ALIGNED DIMENSION command.

20. Select the dimension so it highlights.

Select the two locks on the top dimensions to switch the permanent dimensions to *locked* dimensions. This means these distances will not be changed.

You should see two of the padlocks as closed and one as open.

21. Place an overall dimension using the wall centerlines.

ESC or Cancel out of the ALIGNED DIMENSION command.

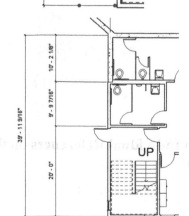

22.

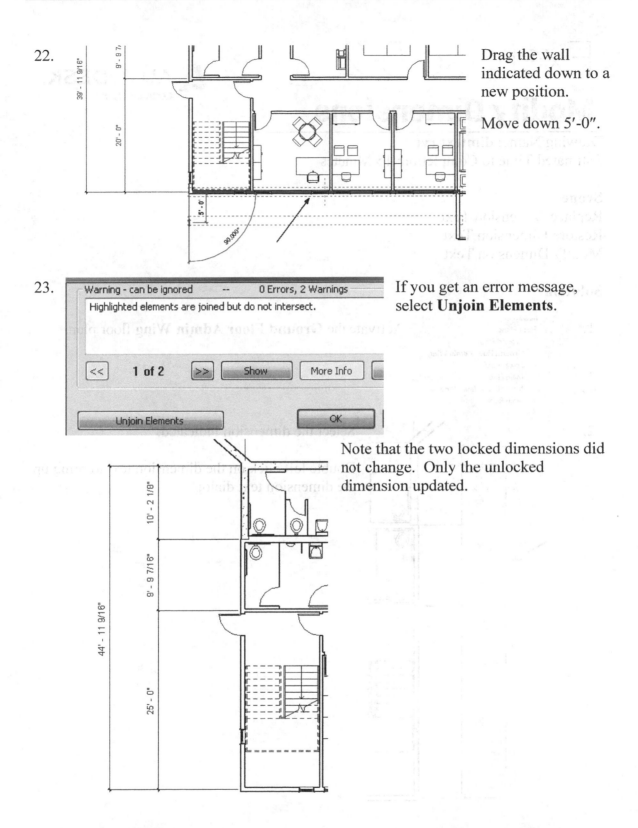

Drag the wall indicated down to a new position.

Move down 5'-0".

23.

If you get an error message, select **Unjoin Elements**.

Note that the two locked dimensions did not change. Only the unlocked dimension updated.

24. Close without saving.

Exercise 5-2
Modify Dimensions

AUTODESK.
Certified User

Drawing Name: **dimtext.rvt**
Estimated Time to Completion: 15 Minutes

Scope
Replace Dimension Text
Restore Dimension Text
Modify Dimension Text

Solution

1. 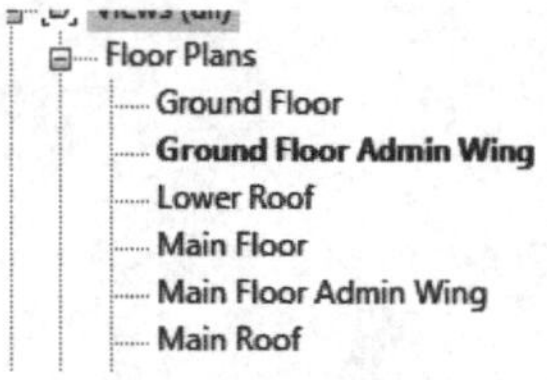 Activate the **Ground Floor Admin Wing** floor plan.

   Views (all)
   Floor Plans
   ── Ground Floor
   Ground Floor Admin Wing
   ── Lower Roof
   ── Main Floor
   ── Main Floor Admin Wing
   ── Main Roof

2. 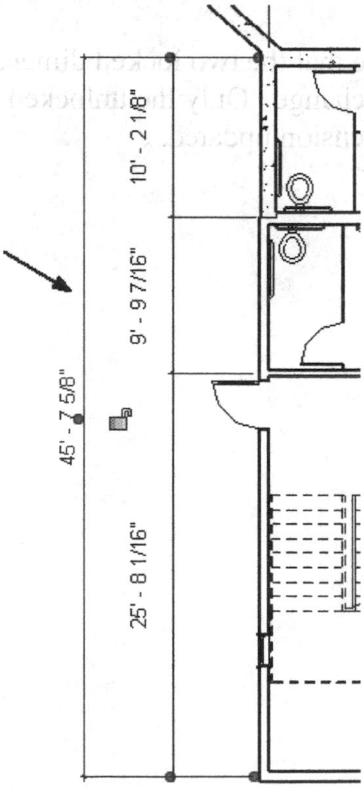 Select the dimension indicated.

   Double left click on the dimension text to bring up the dimension text dialog.

3.
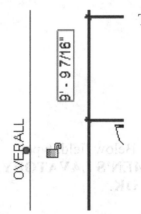
Enable **Replace with Text**.
Enter **45′ 8″**.
Click **OK**.

4.

Invalid Dimension Value

Specify descriptive text for a dimension
segment instead of a numeric value.

To change the dimension value for a length or angle of a
segment, select the element the dimension refers to, and click
the value to edit it.

× You will get an error message stating that
you cannot change the numeric value
without moving the element to correspond
with that value.

Click **Close**.

5.

Enable **Replace with Text**.
Enter **OVERALL**.
Click **OK**.

The dimension text updates.

6. Activate the **Manage** tab on tab on ribbon.
Select **Project Units** on the Settings panel.

Project
rs Units

7.

Discipline:	Common
Units	Format
Angle	12.35°
Area	1235 SF
Cost per Area	[$/ft²] 1235
Distance	1235 [′]
Length	1′ - 5 11/32″
Mass Density	1234.57 lb/ft
Rotation Angle	12.35°

Select the **Length** button under the Format
column.

8.
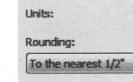

Set the Rounding **To the nearest ½″**.
Click **OK** until all dialogs are closed.

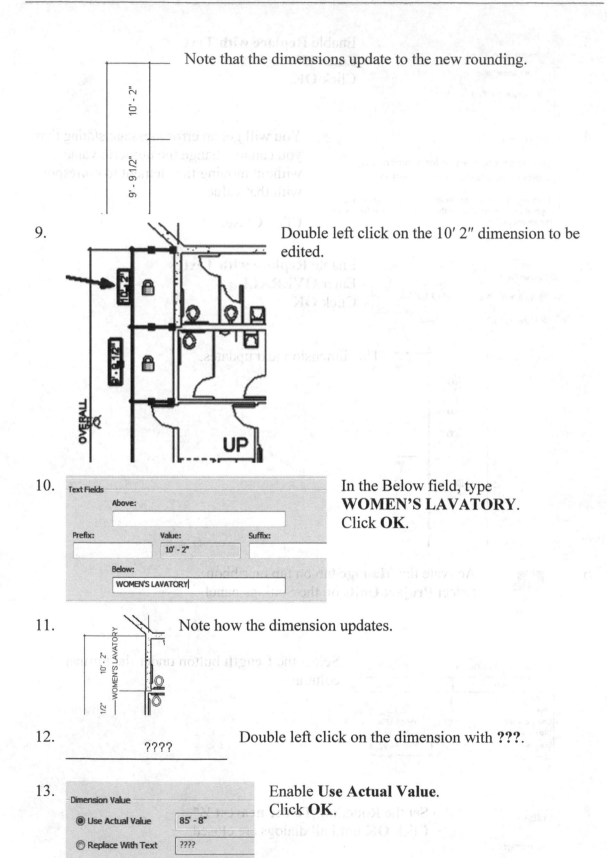

Note that the dimensions update to the new rounding.

9. Double left click on the 10′ 2″ dimension to be edited.

10. In the Below field, type **WOMEN'S LAVATORY**. Click **OK**.

11. Note how the dimension updates.

12. Double left click on the dimension with **???**.

13. Enable **Use Actual Value**. Click **OK**.

14. Close without saving.

Exercise 5-3

Converting Temporary Dimensions to Permanent Dimensions

Drawing Name: **i_dimensions.rvt**
Estimated Time to Completion: 10 Minutes

Scope

Place dimensions using different options
Convert temporary dimension to permanent

Solution

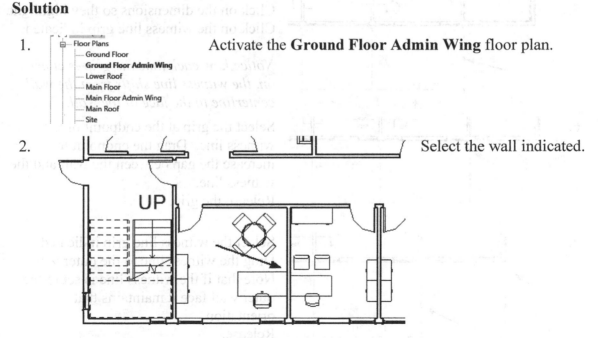

1. Activate the **Ground Floor Admin Wing** floor plan.

   Floor Plans
   —— Ground Floor
   Ground Floor Admin Wing
   —— Lower Roof
   —— Main Floor
   —— Main Floor Admin Wing
   —— Main Roof
   —— Site

2. Select the wall indicated.

3.

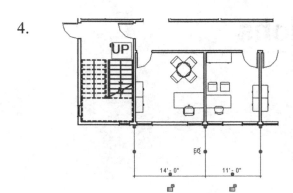

Two dimensions will appear.
There is a small dimension icon visible.
This icon converts a temporary dimension to a permanent dimension.
Left click on this icon.

4.

The dimensions are now converted to permanent dimensions.
Drag the dimensions below the view.

5.

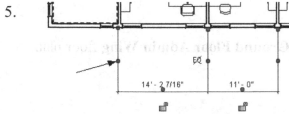

Click on the dimensions so they highlight.
Click on the witness line grip indicated.

Notice how each time the grip is clicked on, the witness line shifts from the wall centerline to the face of the wall.

6.

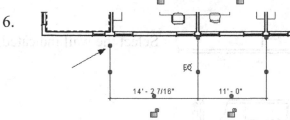

Select the grip at the endpoint of the witness line. Drag the endpoint to increase the gap between the wall and the witness line.
Release the grip.

7.

Select the witness line grip indicated.
Drag the witness line to the outer wall.
Note that if the witness line is set to the outer wall face it maintains that orientation.
Release.

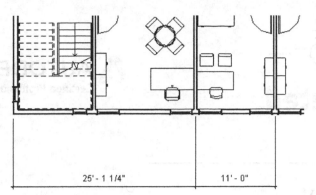

25'- 1 1/4"   11'- 0"

The dimension updates.
Note the new dimension value.

8. Close without saving.

Exercise 5-4
Multi-Segmented Dimensions

Drawing Name: **multisegment.rvt**
Estimated Time to Completion: 10 Minutes

Scope
Add and delete witness lines to a multi-segment dimension

Solution

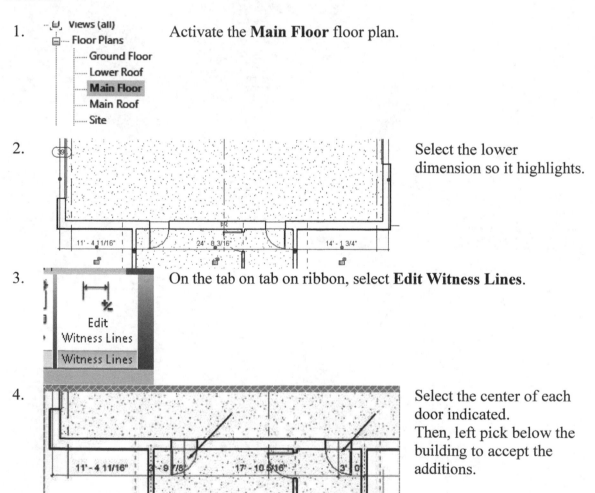

1. Activate the **Main Floor** floor plan.

   Views (all)
   Floor Plans
      Ground Floor
      Lower Roof
      Main Floor
      Main Roof
      Site

2. Select the lower dimension so it highlights.

3. On the tab on tab on ribbon, select **Edit Witness Lines**.

   Edit Witness Lines
   Witness Lines

4. Select the center of each door indicated.
   Then, left pick below the building to accept the additions.

5.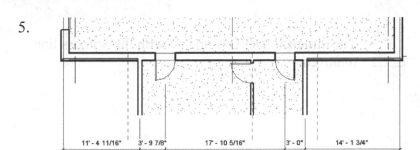

Drag the dimensions down so you can see them easily.

6.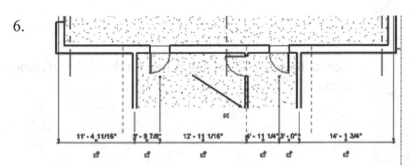

Select the dimension.
Select **Edit Witness Lines**.
Add the witness line at the wall indicated.

7.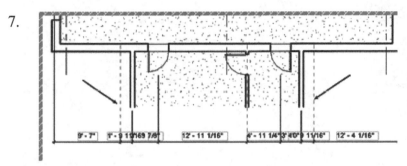

Select the dimension.
Select **Edit Witness Lines**.
Add the witness line at the two reference planes indicated.

Left pick below the building to accept the additions.

8.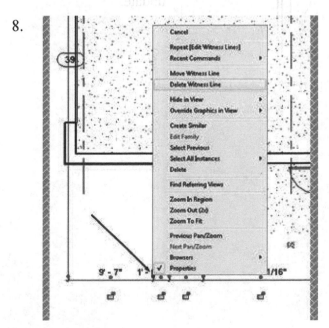

Select the dimension.
Select the grip on the witness line for the left reference plane.
Right click and select **Delete Witness Line**.

9.

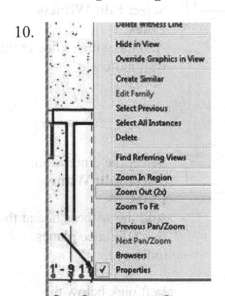

The dimension should update with the witness line removed.

10.

Repeat to delete the witness line on the right reference plane.

11.

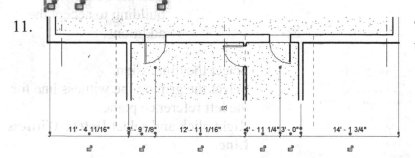

The dimension should update.

12. Close without saving.

Constraints

Revit uses three types of constraints:

- Explicit
- Loose
- Implied

An explicit constraint determines that defined relationships are always maintained. An example of an explicit constraint would be when two elements are aligned and then locked or when a dimension is locked.

A loose constraint is a dimension or alignment which is not locked. They are maintained unless a conflict occurs.

Implied constraints are also maintained unless a conflict occurs, such as when a wall is attached to a roof or two walls are joined at a corner.

An Equality constraint is considered an explicit constraint.

You can display constraints in a view by selecting Reveal Constraints on the View Control bar.

Exercise 5-5
Applying Constraints

Drawing Name: **i_constraints.rvt**
Estimated Time to Completion: 20 Minutes

Scope
Using the Align tool to constrain elements

Solution

1. Activate the **Main Floor** floor plan.

2. Select the interior wall located between Door 25 and Door 26 as indicated.

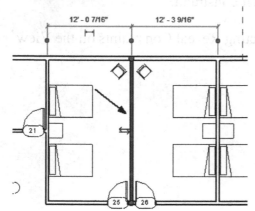

3. The associated temporary dimensions will become visible.
Left click on the permanent dimension toggle to convert the dimensions to permanent dimensions.

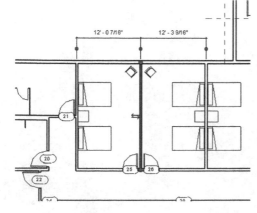

4. Left click anywhere in the display window.
    In the Properties pane,

    scroll down to the **Underlay** parameter.
    Set the Underlay to **Ground Floor**.

    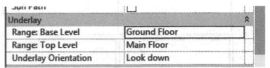

    *This will make the ground floor visible
    in the graphics window. Elements on
    the ground floor will be displayed in a
    lighter shade.*

5.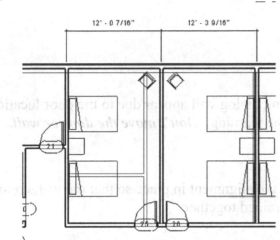

    Note in the upper left corner the interior
    walls are not aligned.

6. Activate the **Modify** tab on tab on ribbon.
    Select the **Align** tool from the Modify panel.

7. Prefer: Wall faces ▾   Set the preference to **Wall faces** on the Options bar.

8.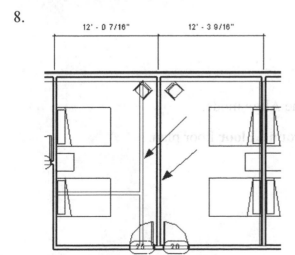

    Select the right side of the ground floor
    interior wall as the source for the alignment.

9.

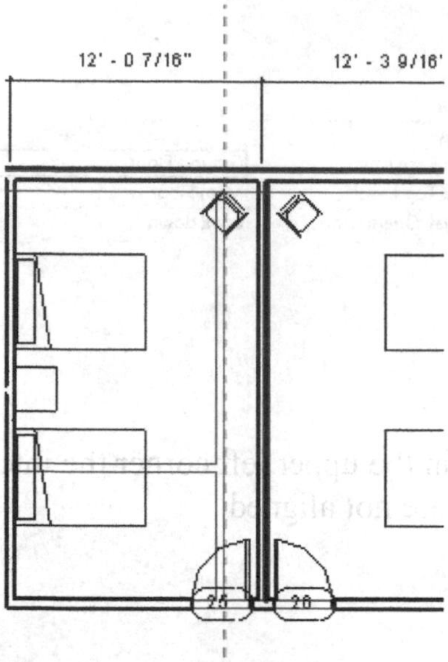

12' - 0 7/16"   12' - 3 9/16'

Select the right side of the main floor interior wall as the target for the alignment (this is the wall that will be shifted).

10.

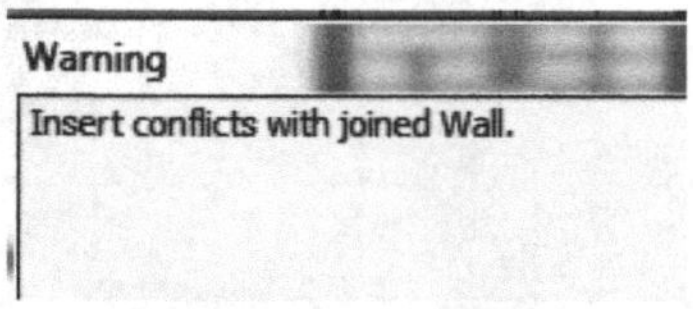

Warning

Insert conflicts with joined Wall.

A warning dialog will appear due to the door location. Ignore the warning. ***Don't move the door or wall.***

11.

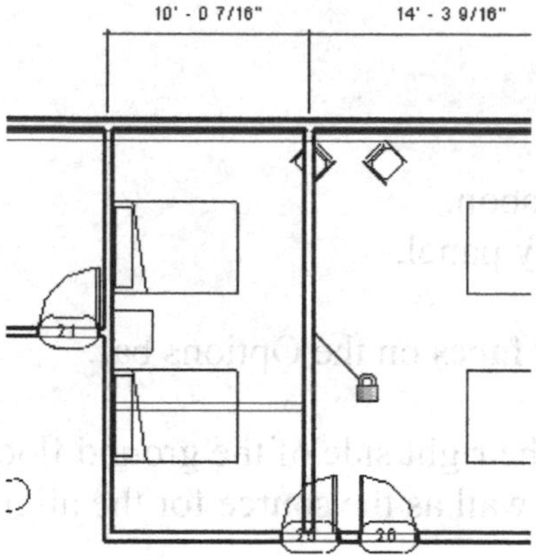

10' - 0 7/16"   14' - 3 9/16"

Lock the alignment in place so that the walls remain constrained together.

12. Right click and select Cancel to exit the Align mode.

13.

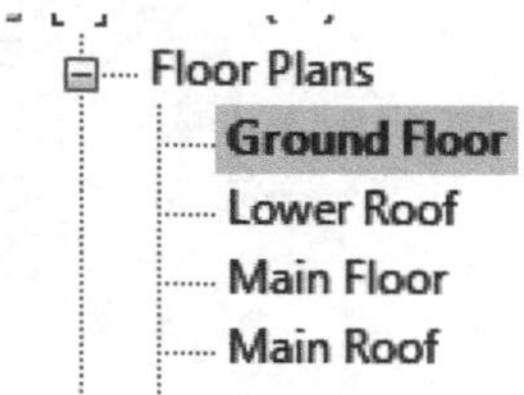

Floor Plans
    Ground Floor
    Lower Roof
    Main Floor
    Main Roof

Activate the **Ground Floor** floor plan.

14. Select the wall, the cabinets, and copier next to the wall.

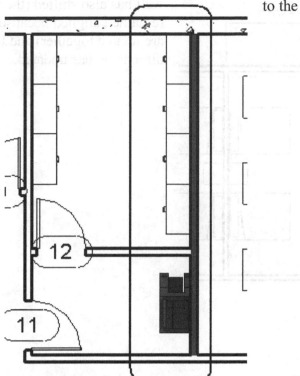

15. Select the **Move** tool on the Modify tab on tab on ribbon.

16. Select a base point.
Move the selected elements to the right **1′ 11″**.
Left click in the window to release the selected elements.

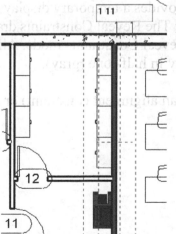

17. Activate the **Main Floor** floor plan.

Floor Plans
  Ground Floor
  Lower Roof
  Main Floor
  Main Roof
  Site
  T. O. Footing
  T. O. Parapet

18.

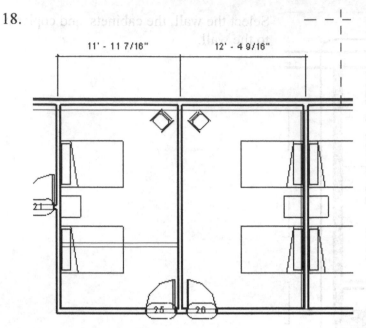

Note that the main floor interior wall has also shifted (the walls remained aligned because they are locked together) and the dimension has updated.

19. Close without saving.

Reveal Constraints

The Reveal Constraints tool on the View Control Bar provides a temporary display of view highlighting dimension and alignment constraints. The Reveal Constraints drawing area displays a color border to indicate that you are in Reveal Constraints mode. All constraints display in color while model elements display in half-tone (gray).

To remove a constraint, you can either unlock it (if it is an alignment constraint) or delete it.

Exercise 5-6
Reveal Constraints

Drawing Name: **reveal_constraints.rvt**
Estimated Time to Completion: 10 Minutes

Scope
Using the Reveal Constraints tool to identify and modify constraints

Solution

1. Floor Plans — Ground Floor — Lower Roof — **Main Floor** — Main Roof   Activate the **Main Floor** floor plan.

2. Click on **Reveal Constraints** on the View Control bar.

    1/8" – 1'-0"

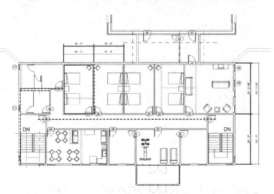

Notice that there are dashed lines as well as dimensions highlighted.

The dashed lines indicate alignment constraints.

3.

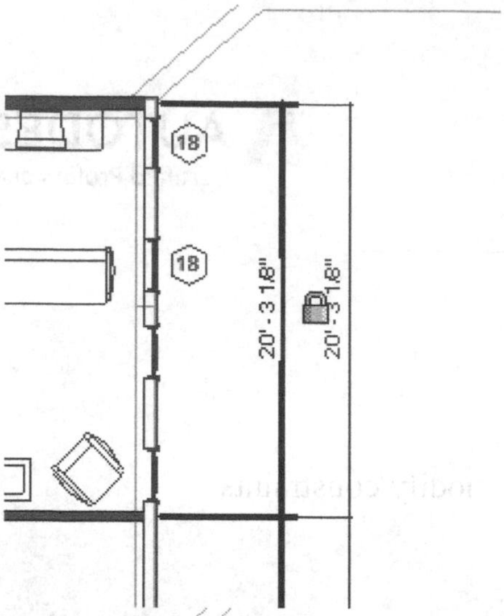

Click on the top vertical dimension on the east side.

Notice this dimension has a lock on it.

Click on the lock to toggle the lock OFF.

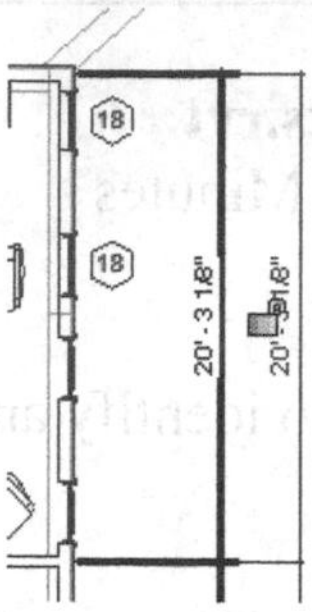

4.

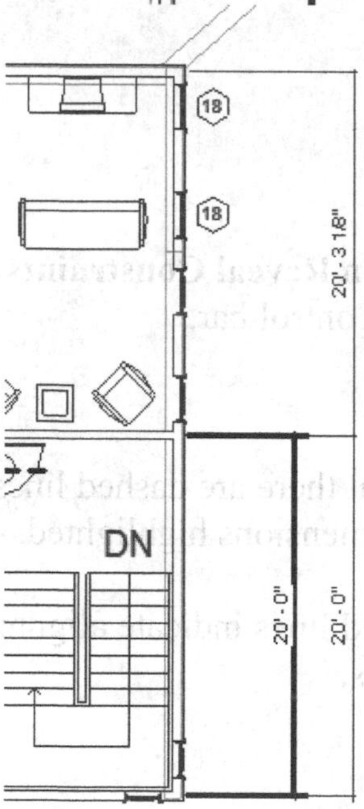

Left click to release the selection.

Notice the unlocked dimension is no longer highlighted.

DN

5.

EQ

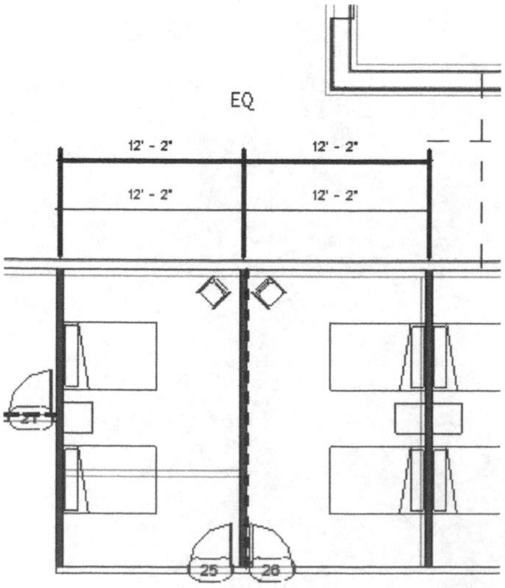

Click on the left horizontal dimension on the North side.

Notice that this dimension has an EQ constraint.

Click on the EQ to disable the constraint.

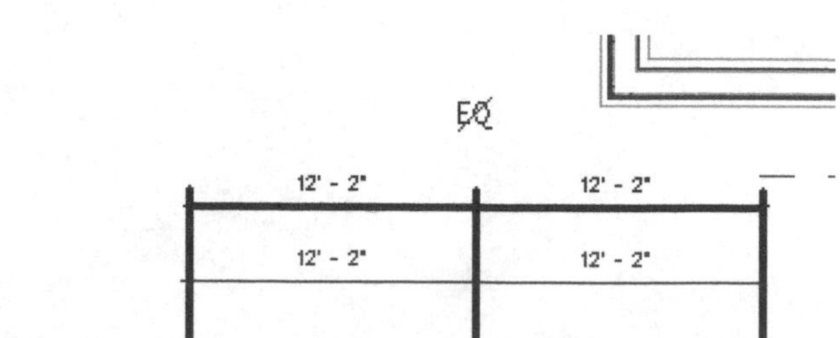

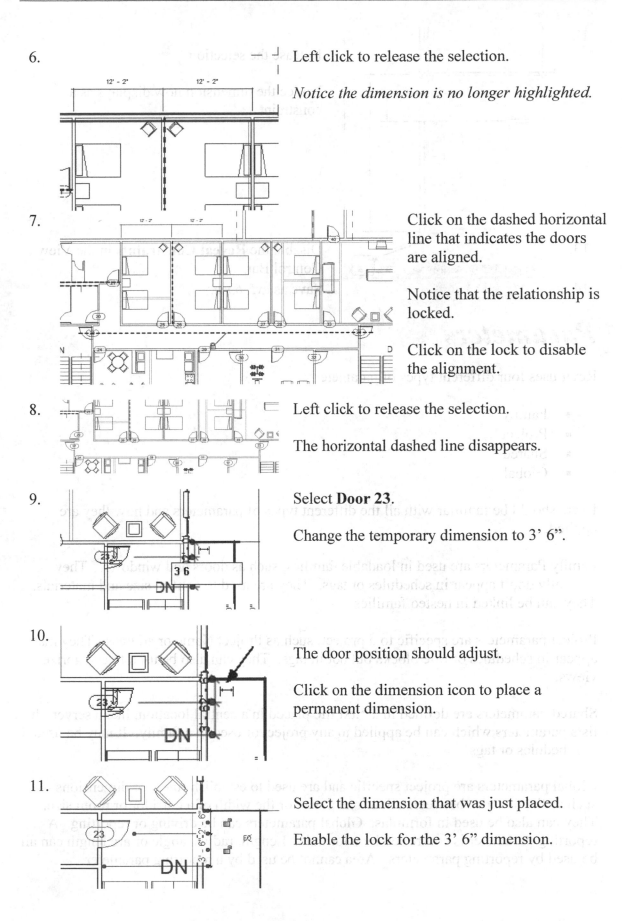

6. Left click to release the selection.

Notice the dimension is no longer highlighted.

7. Click on the dashed horizontal line that indicates the doors are aligned.

Notice that the relationship is locked.

Click on the lock to disable the alignment.

8. Left click to release the selection.

The horizontal dashed line disappears.

9. Select **Door 23**.

Change the temporary dimension to 3' 6".

10. The door position should adjust.

Click on the dimension icon to place a permanent dimension.

11. Select the dimension that was just placed.

Enable the lock for the 3' 6" dimension.

12. 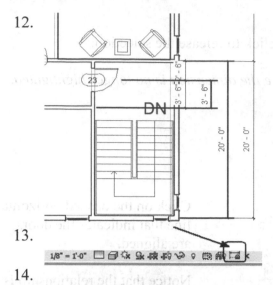 Release the selection.

Notice the dimension now displays as a constraint.

13. Disable the **Reveal Constraints** in the View Control Bar.

14. Save as *ex5-6.rvt*.

Parameters

Revit uses four different types of parameters:

- Family
- Project
- Shared
- Global

Users should be familiar with all the different types of parameters and how they are applied.

Family Parameters are used in loadable families, such as doors and windows. They normally don't appear in schedules or tags. They are used to control size and materials. They can be linked in nested families.

Project parameters are specific to a project, such as Project Name or address. They may appear in schedules or title blocks but not in tags. They can also be used to categorize views.

Shared parameters are defined in a *.txt file placed in a central location, like a server. It lists parameters which can be applied to any project or used in a family. It may be used in schedules or tags.

Global parameters are project specific and are used to establish rules for dimensions – such as the distance between a door and a wall or the width of a corridor or room size. They can also be used in formulas. Global parameters can be driving or reporting. A reporting parameter is controlled by a formula. Length, radius, angle or arc length can all be used by reporting parameters. Area cannot be used by a reporting parameter.

Exercise 5-7

Global Parameters

Drawing Name: **global parameters.rvt**
Estimated Time to Completion: 10 Minutes

Scope
Using global parameters in a project

Solution

1. 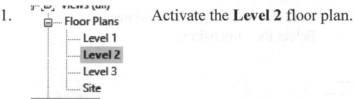 Activate the **Level 2** floor plan.

2. 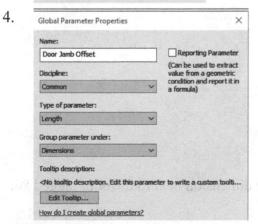 Activate the Manage tab on ribbon.
Select **Global Parameters**.

3. Select **New Global Parameter**.

4. Type **Door Jamb Offse**t.
Click **OK**.

5. Set the value to **1' 0"**.
Click **OK**.

6. 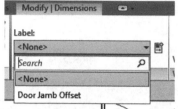 Select the dimension indicating the door jamb offset for the door with the tag labeled 93.

7. 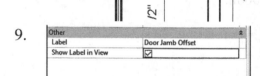 On the ribbon, select **Door Jamb Offset** under Label.

8. Note that the dimension updates.
Select the dimension.

9. 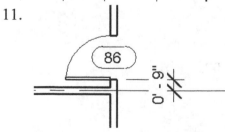 In the Properties panel, enable **Show Label in View**. Click **Apply**.

10. Note that the dimension updates to show the global parameter in the dimension.

11. Select the dimension next to the door tagged 86.

12.
Linear Dimension Style
Linear - 3/32" Arial
Dimensions (1)
Edit Type
Graphics
Leader
Baseline Offset
0"
Text
Value
0' 9"
Other
Label
<None>
<None>
Door Jamb Offset

In the Properties panel,
apply the Door Jamb Offset label.

13.
86
1' - 0"

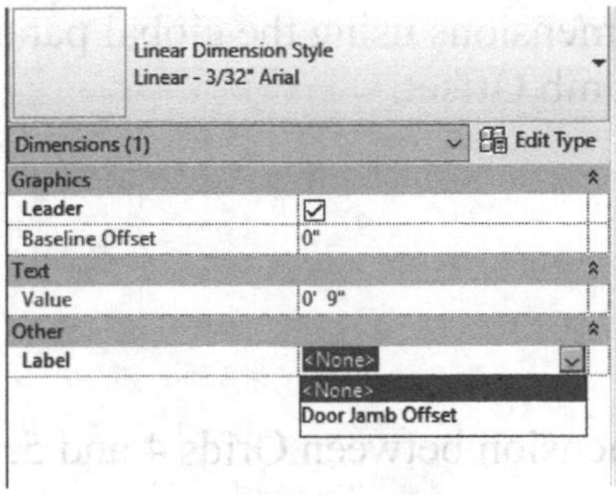

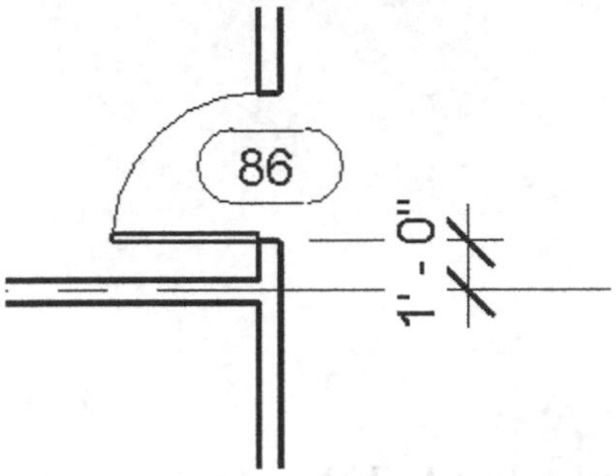

Note that the dimension updates.

14.
82
90
0' - 9"
0' - 9 1/2"

Zoom into the doors tagged 82 and 90.

15.
82
90
1' - 0"
1' - 0"

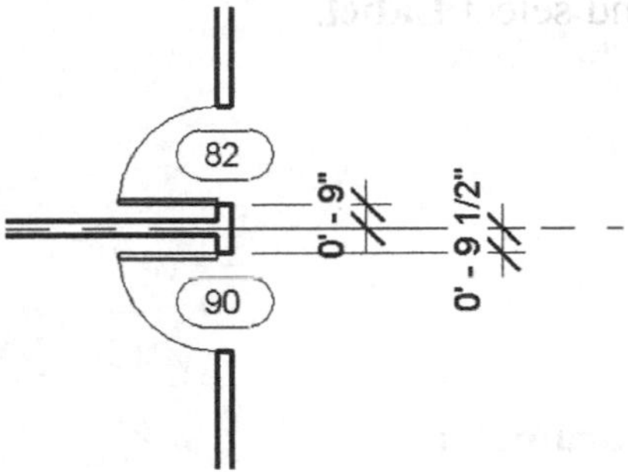

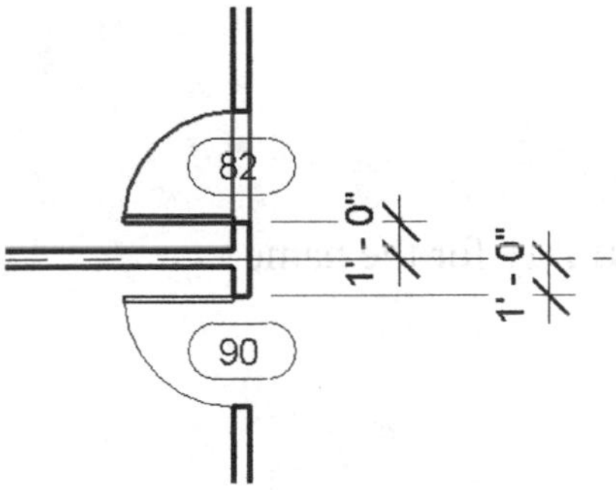

Modify the dimensions using the global parameter called Door Jamb Offset.

You can use the CTL key to select both dimensions and apply the global parameter to both dimensions simultaneously.

16.
91
92
0' - 11"
0' - 7"

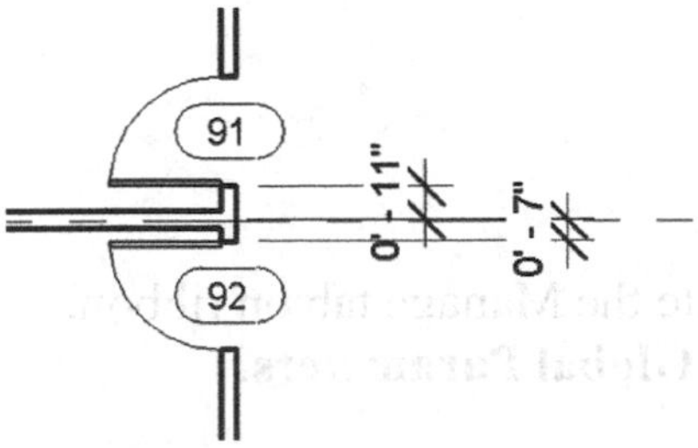

Zoom into the doors tagged 91 and 92.

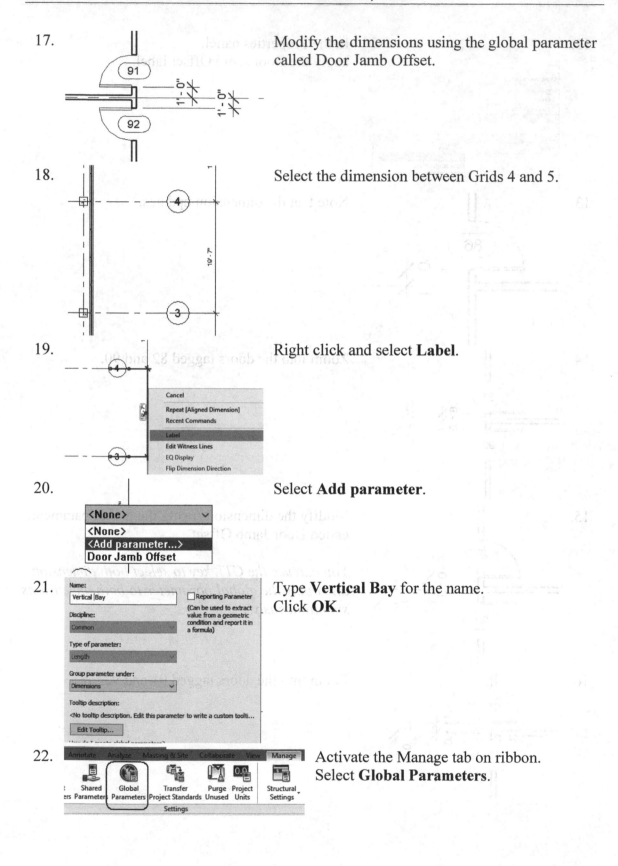

17. Modify the dimensions using the global parameter called Door Jamb Offset.

18. Select the dimension between Grids 4 and 5.

19. Right click and select **Label**.

20. Select **Add parameter**.

21. Type **Vertical Bay** for the name. Click **OK**.

22. Activate the Manage tab on ribbon. Select **Global Parameters**.

23.

Parameter	Value
Dimensions	
Door Jamb Offset	1' 0"
Vertical Bay	18' 2"

Set the value for the Vertical Bay to **18' 2"**.

24.

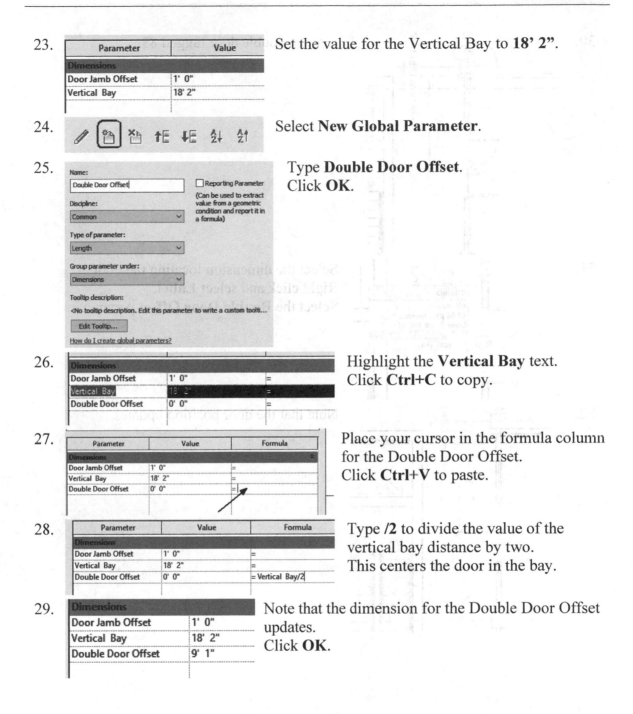

Select **New Global Parameter**.

25.

Name:
Double Door Offset

☐ Reporting Parameter
(Can be used to extract value from a geometric condition and report it in a formula)

Discipline:
Common

Type of parameter:
Length

Group parameter under:
Dimensions

Tooltip description:
<No tooltip description. Edit this parameter to write a custom toolti...

Edit Tooltip...

How do I create global parameters?

Type **Double Door Offset**.
Click **OK**.

26.

Dimensions		
Door Jamb Offset	1' 0"	=
Vertical Bay	18' 2"	=
Double Door Offset	0' 0"	=

Highlight the **Vertical Bay** text.
Click **Ctrl+C** to copy.

27.

Parameter	Value	Formula
Dimensions		
Door Jamb Offset	1' 0"	=
Vertical Bay	18' 2"	=
Double Door Offset	0' 0"	=

Place your cursor in the formula column for the Double Door Offset.
Click **Ctrl+V** to paste.

28.

Parameter	Value	Formula
Dimensions		
Door Jamb Offset	1' 0"	=
Vertical Bay	18' 2"	=
Double Door Offset	0' 0"	= Vertical Bay/2

Type **/2** to divide the value of the vertical bay distance by two.
This centers the door in the bay.

29.

Dimensions	
Door Jamb Offset	1' 0"
Vertical Bay	18' 2"
Double Door Offset	9' 1"

Note that the dimension for the Double Door Offset updates.
Click **OK**.

30. Locate the double door tagged 83.

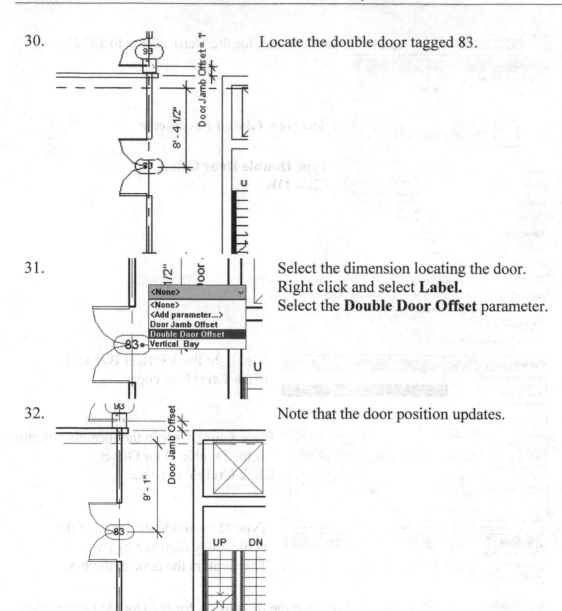

31. Select the dimension locating the door.
Right click and select **Label.**
Select the **Double Door Offset** parameter.

32. Note that the door position updates.

33. Assign the Vertical Bay parameter to the dimensions between grids 2 & 3 and grids 1 & 2.

34. Apply the double door offset parameter to the doors labeled 84 and 85.

35. Use the **ALIGN** tool to move the wall above the door labeled 85 to center it on Grid 2.

Set the Option to **Prefer Wall Centerline**.

36. Lock the wall to the grid.

37. 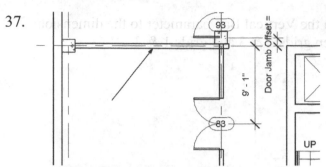 Use the ALIGN tool to position the wall above the door labeled 83 to Grid 4 and lock into position.

    Zoom out so you can see the entire floor plan.

38. Activate the Manage tab on ribbon. Select **Global Parameters**.

39.

Parameter	Value
Dimensions	
Door Jamb Offset	1' 0"
Vertical Bay	18' 6"
Double Door Offset	9' 1"

Change the value of the Vertical Bay to **18' 6"**. Click **OK**.

40. Note that the dimensions update and the double doors remain centered in the bays. Close without saving.

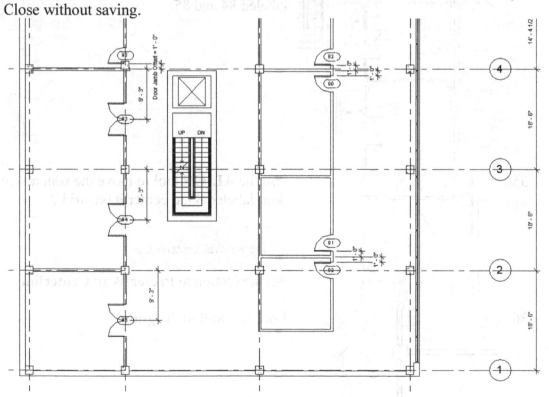

Exercise 5-8
Shared Parameters

Drawing Name: shared_parameters.rvt
Estimated Time: 30 minutes

This exercise reinforces the following skills:

☐ Shared Parameters
☐ Schedules
☐ Family Properties

Many architectural firms have a specific format for schedules. The parameters for these schedules may not be included in the pre-defined parameters in Revit. If you are going to be working in multiple projects, you can define a single file to store parameters to be used in any schedule in any project.

1. **Manage** Activate the **Manage** tab on ribbon.

2. Shared Parameters Select the **Shared Parameters** tool from the Settings panel.

3.  Click **Create** to create a file where you will store your parameters.

4. File name: custom parameters   Locate the folder where you want to store your file.
    Files of type: Shared Parameter Files (*.txt) Set the file name to *custom parameters.txt*.
    Click **Save**.

    Note that this is a txt file.

5. Groups [New...] [Rename...] [Delete] Under Groups, select **New**.

6.  Enter **Door**.
    Click **OK**.

7. Under Parameters, select **New**.

8. Enter **Head Detail** for Name.

    In the Type field, we have a drop-down list. Select **Text**.

    Click **OK**.

9. Notice that we have a Parameter Group called **Door** now.

    There is one parameter listed.

    Select **New** under Parameters.

10. Enter **Jamb Detail** for Name.
    In the Type of Parameter field, select **Text**.
    Click **OK**.

    Select **New** under Parameters.

Hint: *By default, parameters are set to Length. If you rush and don't set the Type of Parameter, you will have to delete it and re-do that parameter.*

11. Enter **Threshold Detail** for Name.
    In the Type field, select **Text**.
    Click **OK**.

    Select **New** under Parameters.

12. Enter **Hardware Group** for Name.
    In the Type field, select **Text**.
    Click **OK**.

13. Select **New** under Groups.

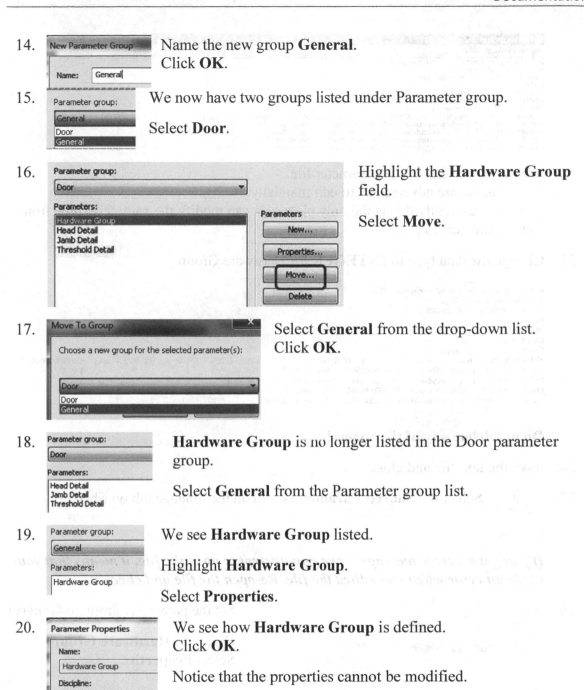

14. Name the new group **General**.
    Click **OK**.

15. We now have two groups listed under Parameter group.

    Select **Door**.

16. Highlight the **Hardware Group** field.

    Select **Move**.

17. Select **General** from the drop-down list.
    Click **OK**.

18. **Hardware Group** is no longer listed in the Door parameter group.

    Select **General** from the Parameter group list.

19. We see **Hardware Group** listed.

    Highlight **Hardware Group**.

    Select **Properties**.

20. We see how **Hardware Group** is defined.
    Click **OK**.

    Notice that the properties cannot be modified.

    Click **OK** to close the dialog.

21. Locate the *custom parameters.txt* file using Windows Explorer.

22. Right click and select **Open**.

This should open the file using Notepad and assumes that you have associated txt files with Notepad.

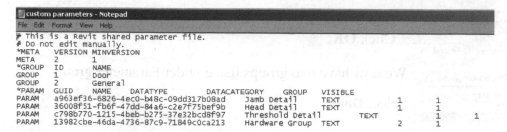

We see the format of the parameter file.

Note that we are advised not to edit manually.

However, currently this is the only place you can modify the parameter type from Text to Integer, etc.

23. Change the data type to **INTEGER** for Hardware Group.

Do not delete any of the spaces!

24. Save the text file and close.

25. 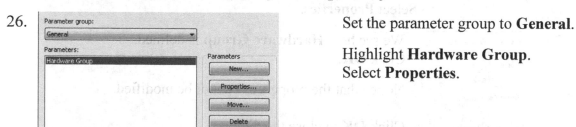 Select the **Shared Parameters** tool on the Manage tab on ribbon.

Shared
Parameters

If you get an error message when you attempt to open the file, it means that you made an error when you edited the file. Re-open the file and check it.

26. Set the parameter group to **General**.

Highlight **Hardware Group**.
Select **Properties**.

27. Note that Hardware Group is now defined as an Integer.

Click **OK** twice to exit the Shared Parameters dialog.

28. Save as *ex5-8.rvt*.

Exercise 5-9

Using Shared Parameters

Drawing Name: shared_parameters.rvt
Estimated Time: 30 minutes

This exercise reinforces the following skills:

- Shared Parameters
- Family Properties
- Schedules

1.  [View]   Activate the **View** tab on ribbon.

2. 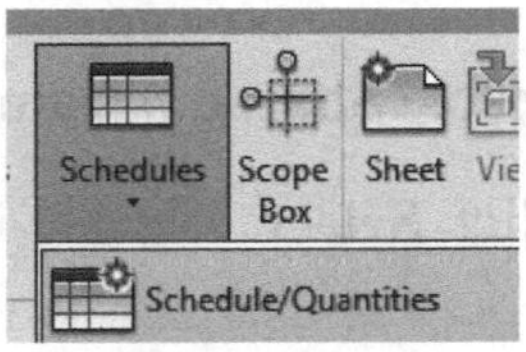   Select **Create→Schedule/ Quantities**.

3. 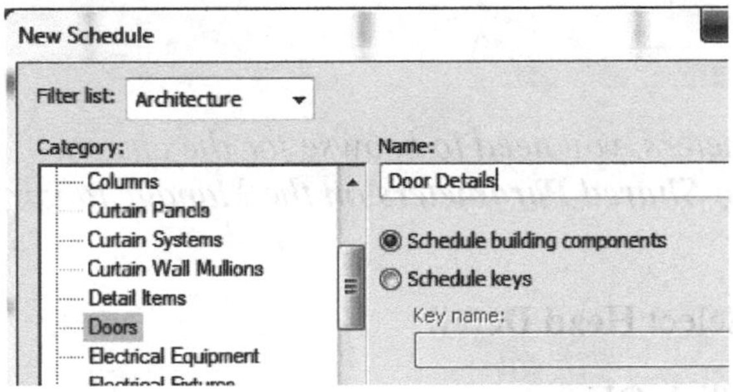   Highlight **Doors**.

    Enter **Door Details** for the schedule name.

    Click **OK**.

    Note that you can create a schedule for each phase of construction.

4.   Add **Mark**, **Type**, **Width**, **Height**, and **Thickness**.

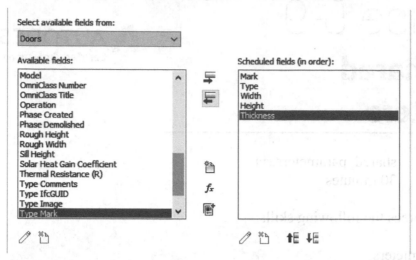

The order of the fields is important. The first field is the first column, etc. You can use the Move Up and Move Down buttons to sort the columns.

5.   Select the **Add Parameter** button.

6.   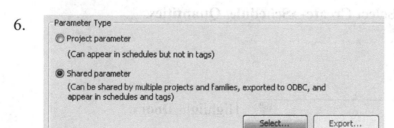   Enable **Shared parameter**.

Click **Select**.

*If you don't see any shared parameters, you need to browse for the custom parameters.txt file to load it. Go to **Shared Parameters** on the Manage tab on ribbon and browse for the file.*

7.   Select **Head Detail**.

Click **OK**.

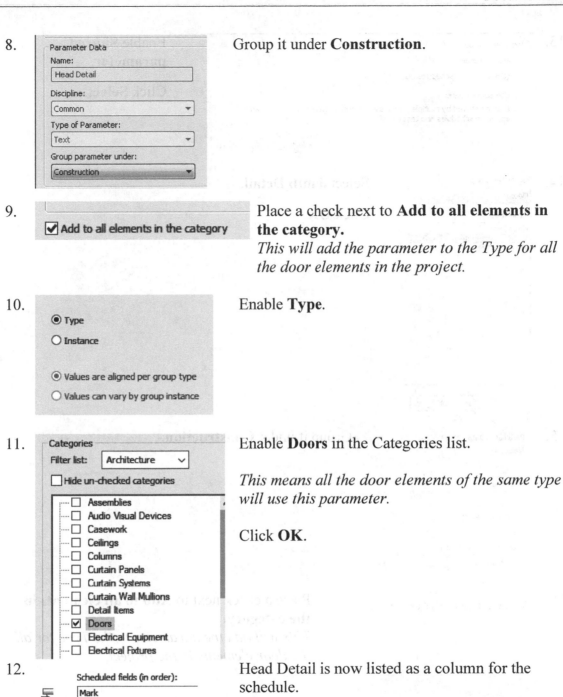

8. Group it under **Construction**.

9. Place a check next to **Add to all elements in the category.**
This will add the parameter to the Type for all the door elements in the project.

10. Enable **Type**.

11. Enable **Doors** in the Categories list.

This means all the door elements of the same type will use this parameter.

Click **OK**.

12. Head Detail is now listed as a column for the schedule.

Select **Add parameter**.

13.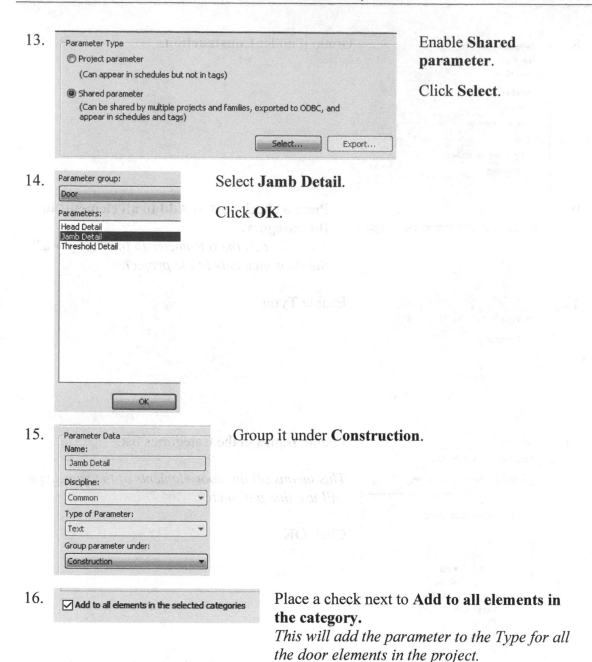

Enable **Shared parameter**.

Click **Select**.

14.

Select **Jamb Detail**.

Click **OK**.

15.

Group it under **Construction**.

16.

Place a check next to **Add to all elements in the category.**
This will add the parameter to the Type for all the door elements in the project.

17. Enable **Doors** in the Categories list.

This means all the door elements of the same type will use this parameter.

Click **OK**.

18. Enable **Type**.

This means all the door elements of the same type will use this parameter.

Click **OK**

19. Jamb Detail is now listed as a column for the schedule.

Select **Add parameter**.

20. Enable **Shared parameter**.

Click **Select**.

21. Select **Threshold Detail**.

Click **OK**.

22.

Parameter Data

Name:
Threshold Detail

Discipline:
Common

Type of Parameter:
Text

Group parameter under:
Construction

Group it under **Construction**.

23.

☑ Add to all elements in the category

Place a check next to **Add to all elements in the category.**
This will add the parameter to the Type Properties for all the door elements in the project.

24.

◉ Type

○ Instance

◉ Values are aligned per group type

○ Values can vary by group instance

Enable **Type**.
This means all the door elements of the same type will use this parameter.

Click **OK**.

Scheduled fields (in order):
Mark
Type
Width
Height
Thickness
Head Detail
Jamb Detail
Threshold Detail

The detail columns have now been added to the schedule.

25.

Heading:
Door No

Heading orientation:
Horizontal

Alignment:
Center

Select the **Formatting** tab.

Highlight **Mark**.
Change the Heading to **Door No**.
Change the Alignment to **Center**.

26.

Heading:
Door Type

Heading orientation:
Horizontal

Alignment:
Left

Change the Heading for Type to **Door Type**.

27.
Heading:
W

Heading orientation:
Horizontal

Alignment:
Center

Change the Heading for Width to **W**.

Change the Alignment to **Center**.

28.
Heading:
H

Heading orientation:
Horizontal

Alignment:
Center

Change the Heading for Height to **H**.

Change the Alignment to **Center**.

29.
Heading:
THK

Heading orientation:
Horizontal

Alignment:
Center

Change the Heading for Thickness to **THK**.

Change the Alignment to **Center**.

Click **OK**.

30. A window will appear with your new schedule.

<Door Details>

A	B	C	D	E	F	G	H
Door No.	Door Type	W	H	THK	Head Detail	Jamb Detail	Threshold Detail
1	36" x 84"	3' - 0"	7' - 0"	0' - 2"			
2	36" x 84"	3' - 0"	7' - 0"	0' - 2"			
3	36" x 84"	3' - 0"	7' - 0"	0' - 2"			
4	36" x 84"	3' - 0"	7' - 0"	0' - 2"			
5	72" x 82"	6' - 0"	6' - 10"	0' - 1 3/4"			
6	72" x 82"	6' - 0"	6' - 10"	0' - 1 3/4"			
7	36" x 84"	3' - 0"	7' - 0"	0' - 2"			
8	36" x 84"	3' - 0"	7' - 0"	0' - 2"			
9	36" x 84"	3' - 0"	7' - 0"	0' - 2"			
10	36" x 84"	3' - 0"	7' - 0"	0' - 2"			
11	12000 x 2290 Ope	2' - 11 1/2"	7' - 6 1/4"	0' - 1 1/4"			
12	12000 x 2290 Ope	2' - 11 1/2"	7' - 6 1/4"	0' - 1 1/4"			
21	Door-Curtain-Wall-	6' - 9"	8' - 0"				

31.

C	D	E
W	H	THK

Select the W column, then drag your mouse to the right to highlight the H and THK columns.

32.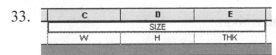
Group

Select **Titles & Headers→Group** on the ribbon.

33.

C	D	E
	SIZE	
W	H	THK

Type '**SIZE**' as the header for the three columns.

34.

F	G	H
Head Detail	Jamb Detail	Threshold Detail

Select the Head Detail column, then drag your mouse to the right to highlight the Jamb Detail and Threshold Detail columns.

35. [icon] **Group**   Select **Titles & Headers→Group**.

36.

F	G	H
	DETAILS	
Head Detail	Jamb Detail	Threshold Detail

Type '**DETAILS**' as the header for the three columns.

37. Our schedule now appears in the desired format.

	3D-Lobby		Level 1		View 1		Level 2		Door Schedule		Door Details X

			<Door Details>				

A	B	C	D	E	F	G	H
			SIZE			DETAILS	
Door No	Door Type	W	H	THK	Head Detail	Jamb Detail	Threshold Detail
1	36" x 84"	3' - 0"	7' - 0"	0' - 2"			
2	36" x 84"	3' - 0"	7' - 0"	0' - 2"			
3	36" x 84"	3' - 0"	7' - 0"	0' - 2"			
5	36" x 84"	3' - 0"	7' - 0"	0' - 2"			
6	72" x 82"	6' - 0"	6' - 10"	0' - 1 3/4"			
7	72" x 82"	6' - 0"	6' - 10"	0' - 1 3/4"			
8	36" x 84"	3' - 0"	7' - 0"	0' - 2"			
9	36" x 84"	3' - 0"	7' - 0"	0' - 2"			
10	36" x 84"	3' - 0"	7' - 0"	0' - 2"			
11	36" x 84"	3' - 0"	7' - 0"	0' - 2"			
12	12000 x 2290 Ope	2' - 11 7/16"	7' - 6 5/32"	0' - 1 3/16"			
13	12000 x 2290 Ope	2' - 11 7/16"	7' - 6 5/32"	0' - 1 3/16"			
21	Door-Curtain-Wall-	6' - 9 1/16"	8' - 0"				

38. Save as *ex5-9.rvt*.

Exercise 5-10
Project Parameters

A AUTODESK.
Certified Professional

Drawing Name: *project_parameters.rvt*
Estimated Time: 20 minutes

This exercise reinforces the following skills:

- Project Browser
- Project Parameters

1.   [**Manage**]   Select the **Manage** tab on ribbon.

2.   [icon] Project Parameters   Select the **Project Parameters** tool.

3.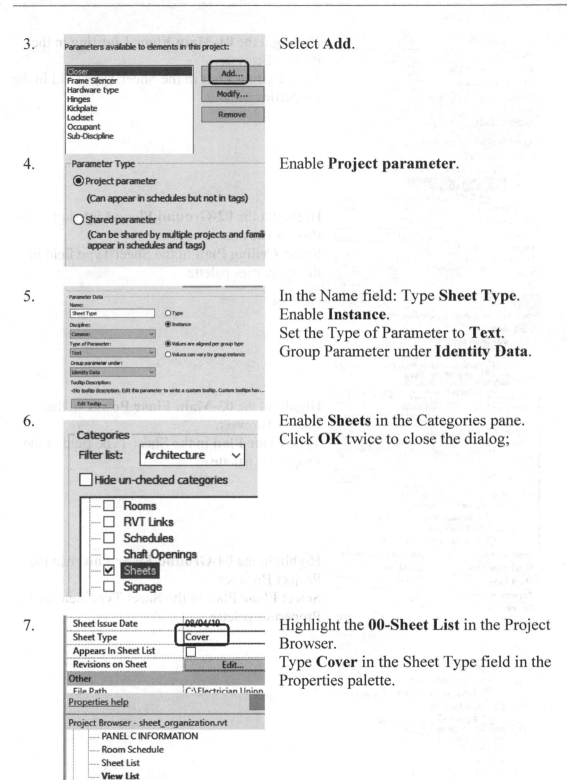

Select **Add**.

4. Enable **Project parameter**.

5. In the Name field: Type **Sheet Type**.
Enable **Instance**.
Set the Type of Parameter to **Text**.
Group Parameter under **Identity Data**.

6. Enable **Sheets** in the Categories pane.
Click **OK** twice to close the dialog;

7. Highlight the **00-Sheet List** in the Project Browser.
Type **Cover** in the Sheet Type field in the Properties palette.

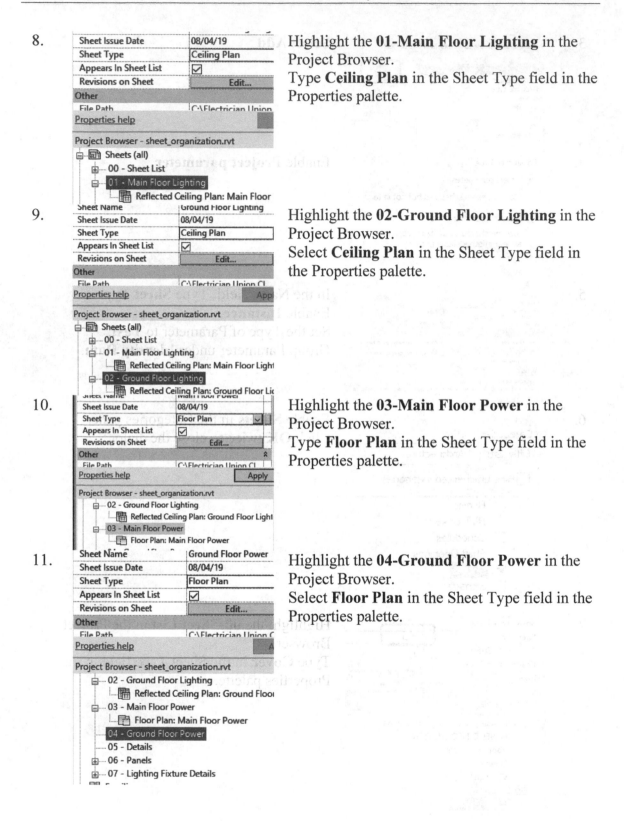

8. Highlight the **01-Main Floor Lighting** in the Project Browser.
Type **Ceiling Plan** in the Sheet Type field in the Properties palette.

9. Highlight the **02-Ground Floor Lighting** in the Project Browser.
Select **Ceiling Plan** in the Sheet Type field in the Properties palette.

10. Highlight the **03-Main Floor Power** in the Project Browser.
Type **Floor Plan** in the Sheet Type field in the Properties palette.

11. Highlight the **04-Ground Floor Power** in the Project Browser.
Select **Floor Plan** in the Sheet Type field in the Properties palette.

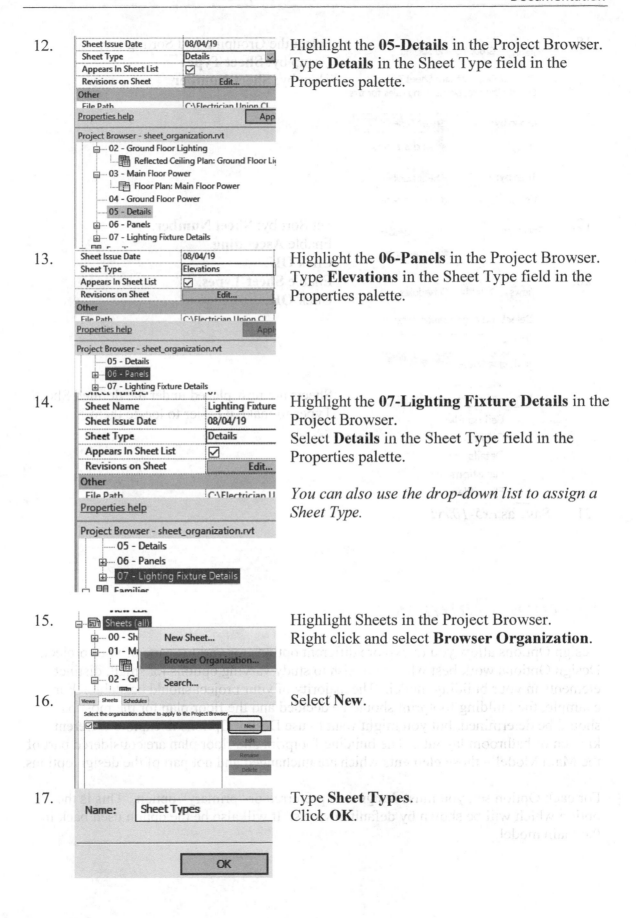

12. Highlight the **05-Details** in the Project Browser. Type **Details** in the Sheet Type field in the Properties palette.

13. Highlight the **06-Panels** in the Project Browser. Type **Elevations** in the Sheet Type field in the Properties palette.

14. Highlight the **07-Lighting Fixture Details** in the Project Browser.
Select **Details** in the Sheet Type field in the Properties palette.

You can also use the drop-down list to assign a Sheet Type.

15. Highlight Sheets in the Project Browser.
Right click and select **Browser Organization**.

16. Select **New**.

17. Type **Sheet Types**.
Click **OK**.

18. 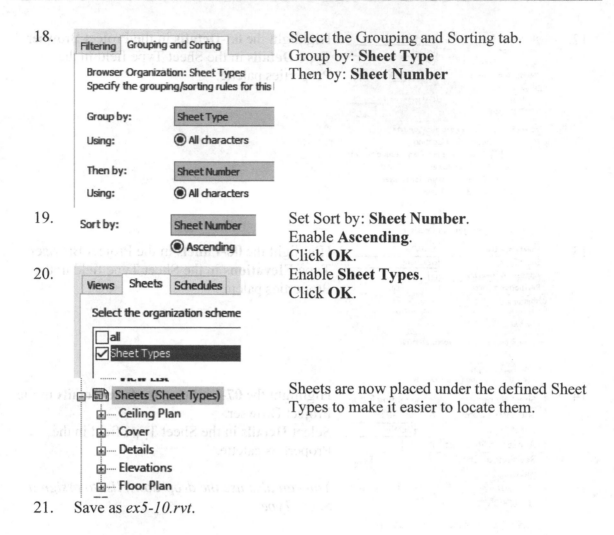 Select the Grouping and Sorting tab.
Group by: **Sheet Type**
Then by: **Sheet Number**

19. Set Sort by: **Sheet Number**.
Enable **Ascending**.
Click **OK**.

20. Enable **Sheet Types**.
Click **OK**.

Sheets are now placed under the defined Sheet Types to make it easier to locate them.

21. Save as *ex5-10.rvt*.

Design Options

Design Options allow you to explore different options for various parts of your project. Design Options work best when you wish to study varying options for small, distinct elements in your building model. The majority of your project should be stable. For example, the building footprint should be decided and the floor plan for the most part should be determined, but you might want to use Design Options to explore different kitchen or bathroom layouts. The building footprint and floor plan are considered part of the Main Model – those elements which are unchanged and not part of the design options.

For each Option set, you must designate a preferred or "primary" option. This is the option which will be shown by default in views. It will also be the option used back in the main model.

Design Options are considered an advanced tool, so you won't be asked about them for the User exam, but you will be expected to answer at least one question regarding Design Options for the Professional exam.

Exercise 5-11

Design Options

Drawing Name: **i_Design_Options**
Estimated Time to Completion: 90 Minutes

Scope
Use of Design Options
Place Views on Sheets
Visibility/Graphics Overrides
Duplicate Views
Rename Views

Solution

1. 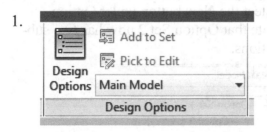 Activate the **Manage** tab on tab on ribbon. Select **Design Options** under the Design Options panel.

2. 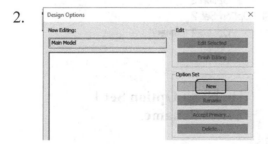 Select **New** under Option Set. Select **New** a second time.

3.
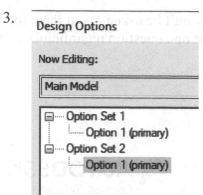

There should be two Option Sets displayed in the left panel.
Each Option set represents a design choice group. The Option set can have as many options as needed. The more options, the larger your file size will become.

4.
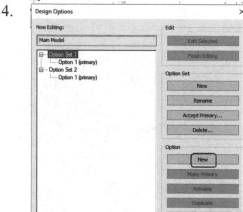

Highlight the **Option Set 1**.
Select the **New** button under Option.
Note that Option Set 1 now has two sub-options.

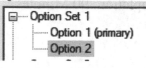

5.
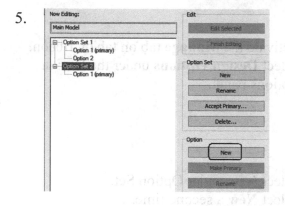

Highlight the **Option Set 2**.
Select the **New** button under Option.
Note that Option Set 2 now has two sub-options.

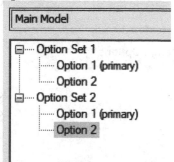

6.
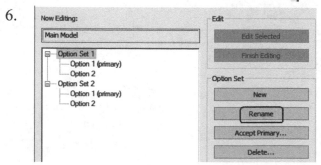

Highlight **Option Set 1**.
Select **Rename**.

7. 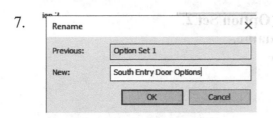 Rename Option Set 1 **South Entry Door Options**.
Click **OK**.

8. 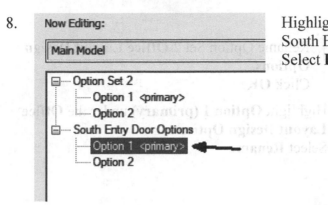 Highlight **Option 1 (primary)** under the South Entry Door Options.
Select **Rename**.

9. 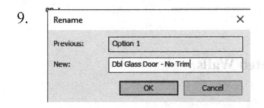 Rename to **Dbl Glass Door - No Trim**.
Click **OK**.

10. 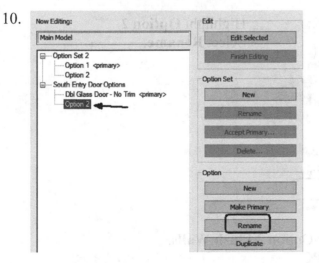 Highlight **Option 2** under the South Entry Door Options.
Select **Rename**.

11. 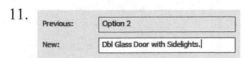 Rename to **Dbl Glass Door with Sidelights**.
Click **OK**.

12.

Now Editing:

Main Model

- Option Set 2
  - Option 1 <primary>
  - Option 2
- South Entry Door Options
  - Dbl Glass Door - No Trim <primary>
  - Dbl Glass Door with Sidelights

Highlight **Option Set 2**.
Select **Rename**.

13.

Previous: Option Set 2

New: Office Layout Design Options

Rename Option Set 2 **Office Layout Design Options**.
Click **OK**.

14.

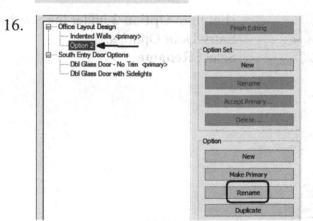

Highlight **Option 1 (primary)** under the **Office Layout Design Options**.
Select **Rename**.

15.

Previous: Option 1

New: Indented Walls

Rename to **Indented Walls**.
Click **OK**.

16.

- Office Layout Design
  - Indented Walls <primary>
  - Option 2
- South Entry Door Options
  - Dbl Glass Door - No Trim <primary>
  - Dbl Glass Door with Sidelights

Finish Editing

Option Set

New

Rename

Accept Primary...

Delete...

Option

New

Make Primary

Rename

Duplicate

Highlight **Option 2**.
Select **Rename**.

17.

Previous: Option 2

New: Flush Walls

Rename Option 2 **Flush Walls**.
Click **OK**.

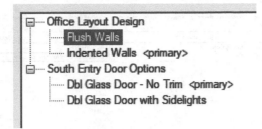

You should have two Option Sets.
Each Option Set should have two options.

Notice that Revit automatically sorted the Options and the sets alphabetically.

An Option Set can have as many options as you like, but the more option sets and options, the larger your file size and the more difficult it becomes to manage.

18. Close the Design Options dialog.

19. 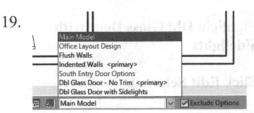   Note in the bottom of the window, you can select which Option set you want active.

20. Using **Duplicate View→Duplicate**, create four copies of the Level 1 view.

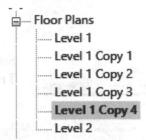

Rename the duplicate views:
Level 1 - Office Layout Indented Walls
Level 1 - Office Layout Flush Walls
Level 1 - South Entry Dbl Glass Door – No Trim
Level 1 - South Entry Dbl Glass Door with Sidelights

To rename, highlight the level name and Click F2.

21. 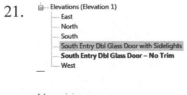   Using **Duplicate View→Duplicate**, create two copies of the South Elevation view.
Rename the duplicate views:
South Entry Dbl Glass Door – No Trim
South Entry Dbl Glass Door with Sidelights

22. Activate **Level 1 - South Entry Dbl Glass Door – No Trim**.

23. | Parts Visibility | Show Original |
    | Visibility/Graphics Overrides | Edit... |
    | Graphic Display Options | Edit... |
    | Underlay | None |

In the Properties pane:
Select **Edit** Visibilities/Graphics Overrides.

24. Activate the **Design Options** tab.

    Set **Dbl Glass Door – No Trim** on South Entry Door Options.
    Click **Apply** and **OK**.

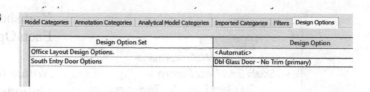

Design Option Set	Design Option
Office Layout Design Options.	<Automatic>
South Entry Door Options	Dbl Glass Door - No Trim (primary)

25. Select the **Design Options** button.

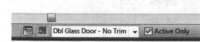

26.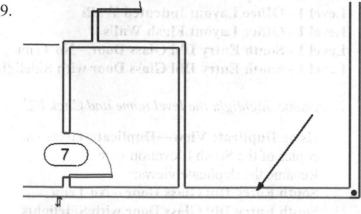

    Highlight **Dbl Glass Door with Sidelights**.

    Click **Edit Selected**.

    Click **Close**.

27. Set the Design Option to **Dbl Glass Door - No Trim (primary)**.

28. Uncheck **Active Only**.

29. Select the south horizontal wall.

30. Activate the **Manage** tab on tab on ribbon.
    Under Design Options, select **Add to Set**.
    *The selected wall is added to the **Dbl Glass Door - No Trim (primary)** set.*
    We need to add the wall to the set so we can place a door. Remember doors are wall-hosted.

31. 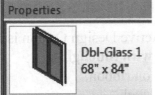 Activate the **Architecture** tab on tab on ribbon. Select the **Door** tool from the Build panel.

32. 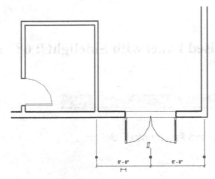 Set the Door type to **Dbl-Glass 1: 68″ x 84″**.

33. Place the door as shown.

34. Select the **Design Options** button.

35. Click **Finish Editing**.

    Click **Close**.

36. - Floor Plans
    - Level 1
    - Level 1 – Office Layout Flush Walls
    - Level 1 – Office Layout Indented Walls
    - **Level 1 – South Entry Dbl Glass Door with Sidelights**
    - Level 1 – South Entry Dbl Glass Door – No Trim
    - Level 2

    Activate **Level 1 – South Entry Dbl Glass Door with Sidelights**.

37. | Parts Visibility | Show Original |
    |---|---|
    | Visibility/Graphics Overrides | Edit... |
    | Graphic Display Options | Edit... |
    | Underlay | None |

    In the Properties pane:
    Select **Edit** Visibilities/Graphics Overrides.

38. Activate the **Design Options** tab.

Design Option Set	Design Option
South Entry Door Options	Dbl Glass Door with Sidelights
Office Layout Design Options	<Automatic>

    Set **Dbl Glass Door with Sidelights** on South Entry Door Options.
    Click **OK**.

39. 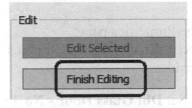 Select the **Design Options** button.

40. 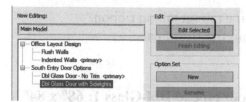 Highlight **Dbl Glass Door with Sidelights**.

Click **Edit Selected**.

Close the dialog.

41.  Verify the Active Design Option is **Dbl Glass Door with Sidelights.**

42. Activate the **Architecture** tab on tab on ribbon.
Select the **Door** tool from the Build panel.

43. 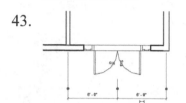 Place a **Double-Raised Panel with Sidelights: 68″ x 80″** door as shown.

44. Activate the **South Entry Dbl Glass Door – No Trim** elevation.

45. In the Properties pane:
Select **Edit** Visibilities/Graphics Overrides.

46. Activate the **Design Options** tab.

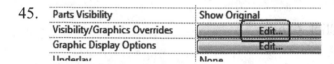

Set **Dbl Glass Door - No Trim** on South Entry Door Options.
Click **OK**.

47. Activate the **South Entry Dbl Glass Door with Sidelights** elevation.

48. In the Properties pane:
Select **Edit** Visibilities/Graphics Overrides.

49. Activate the **Design Options** tab.

Set **Dbl Glass Door with Sidelights** on South Entry Door Options.

Click **OK**.

50.
Sheets (all)
A101 - South Entry Door Option
A102 - Office Layout Options

Activate the Sheet named **South Entry Door Options**.

51.

Drag and drop the two South Entry Option elevation views on the sheet.

Level 2
10' - 0"

Level 1
0' - 0"

① South Entry Dbl Glass Door with Sidelights
1/8" = 1'-0"

Level 2
10' - 0"

Level 1
0' - 0"

② South Entry Dbl Glass Door - No Trim
1/8" = 1'-0"

52.

Switch to 3D view.

53. Use **Duplicate View→Duplicate** to create two new 3D views.

3D Views
3D - South Entry Dbl Glass Door - No Trim
3D - South Entry Dbl Glass Door with Sidelights
{3D}

Rename the views:
3D - South Entry Dbl Glass Door - No Trim
3D - South Entry Dbl Glass Door with Sidelights

54.
3D Views
3D - South Entry Option 1
3D - South Entry Option 2
{3D}

Activate **3D - South Entry Dbl Glass Door - No Trim**.

55.
Detail Level	Coarse
Visibility/Graphics Overrides	Edit...
Visual Style	Hidden Line
Graphic Display Options	Edit...

In the Properties pane:
Select **Edit** Visibilities/Graphics Overrides.

56.
Design Option Set	
South Entry Door Options	Dbl Glass Door - No Trim (primary)
Office Layout Design Options	<Automatic>

Activate the **Design Options** tab.

Set **Dbl Glass Door - No Trim** on South Entry Door Options.
Click **OK**.

57.
Model Categories | Annotation Categories | Analytical Model Categories | Imported Categories
☑ Show annotation categories in this view
Filter list: Architecture ⌄

Visibility	Projection/Surface Lines	Halftone
☑ Furniture Tags		☐
☑ Generic Annotations		☐
☑ Generic Model Tags		☐
☑ Grids		☐
☑ Guide Grid		
☑ Keynote Tags		☐
☐ Levels		☐
☑ Lighting Fixture Tags		☐

Disable the visibility of **LEVELS** on the Annotation Categories tab.

58.
3D Views
3D - South Entry Dbl Glass Door - No Trim
3D - South Entry Dbl Glass Door with Sidelights
{3D}

Activate **3D - South Entry Dbl Glass Door with Sidelights**.

59.

Display Model	Normal
Detail Level	Coarse
Visibility/Graphics Overrides	Edit...
Visual Style	Hidden Line
Graphic Display Options	Edit...

In the Properties pane:
Select **Edit** Visibilities/Graphics Overrides.

60.

Visibility/Graphic Overrides for 3D View: 3D - South Entry Dbl Glass Door with Sidelights

| Model Categories | Annotation Categories | Analytical Model Categories | Imported Categories | Filters |

Design Option Set	Design Option
Office Layout Design	<Automatic>
South Entry Door Options	Dbl Glass Door with Sidelights

Activate the **Design Options** tab.

Set Dbl Glass Door with Sidelights on South Entry Door Options.
Click **OK**.

If you forget which design option you are working in, check the title at the top of the dialog.

61.

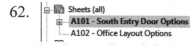

| Model Categories | Annotation Categories | Analytical Model Categories | Imported Categories |

☑ Show annotation categories in this view

Filter list: Architecture

Visibility	Projection/Surface Lines	Halftone
☑ Furniture Tags		☐
☑ Generic Annotations		☐
☑ Generic Model Tags		☐
☑ Grids		☐
☑ Guide Grid		
☑ Keynote Tags		☐
☐ Levels		☐
☑ Lighting Fixture Tags		☐

Disable the visibility of **LEVELS** on the Annotation Categories tab.

62.

🗐 Sheets (all)
  ⊞ **A101 - South Entry Door Options**
  A102 - Office Layout Options

Activate the Sheet named **South Entry Door Options**.

63.

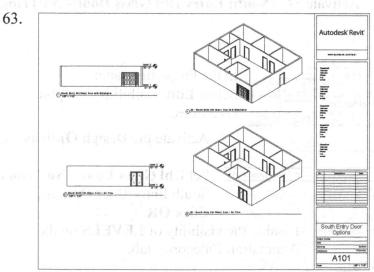

Drag and drop the 3D views onto the sheet.

64.

Select the **Design Options** button.

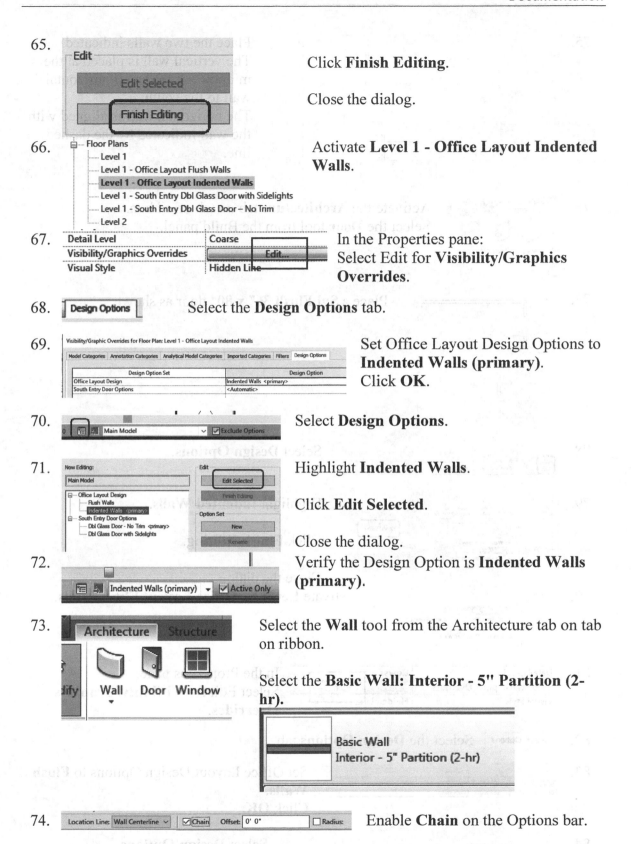

65. Click **Finish Editing**.

    Close the dialog.

66. Activate **Level 1 - Office Layout Indented Walls**.

67. In the Properties pane:
    Select Edit for **Visibility/Graphics Overrides**.

68. Select the **Design Options** tab.

69. Set Office Layout Design Options to **Indented Walls (primary)**.
    Click **OK**.

70. Select **Design Options**.

71. Highlight **Indented Walls**.

    Click **Edit Selected**.

    Close the dialog.

72. Verify the Design Option is **Indented Walls (primary)**.

73. Select the **Wall** tool from the Architecture tab on tab on ribbon.

    Select the **Basic Wall: Interior - 5" Partition (2-hr).**

74. Enable **Chain** on the Options bar.

75. 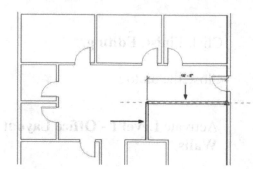 Place the two walls indicated. The vertical wall is placed at the midpoint of the small horizontal wall to the south. The horizontal wall is aligned with the wall indicated by the dashed line.

76. Activate the **Architecture** tab on tab on ribbon. Select the **Door** tool from the Build panel.

77. 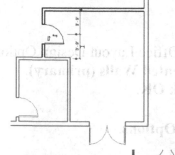 Place a **Sgl Flush 36″ x 80″** door as shown.

78. Select **Design Options**.

79. Highlight **Indented Walls**.

   Click **Finish Editing**.

   Close the dialog.
80. Activate **Level 1 - Office Layout Flush Walls**.

81. In the Properties pane: Select Edit for **Visibility/Graphics Overrides**.

82. Select the **Design Options** tab.

83. Set Office Layout Design Options to **Flush Walls**. Click **OK**.

84.  Select **Design Options**.

85.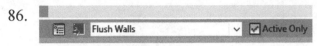

Highlight **Flush Walls**.

Click **Edit Selected**.

Close the dialog.

86.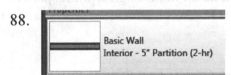

Verify the Design Option is **Flush Walls**.

87. Activate the **Architecture** tab on tab on ribbon.
Select the **Wall** tool from the Build panel.

88. On the Properties pane:
Select the **Basic Wall: Interior - 5″ Partition (2 hr)** wall type.

89.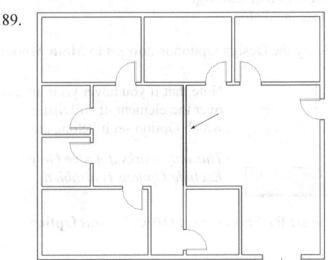

Add the wall shown.

Note that the walls and door added for the Indented Walls option are not displayed.

90.

Activate the **Architecture** tab on tab on ribbon.
Select the **Door** tool from the Build panel.

91. 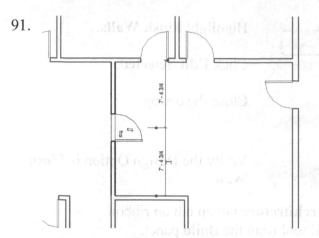 Place a **Sgl Flush 36″ x 84″** door as shown.

92. 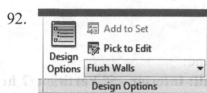 Activate the **Manage** tab on tab on ribbon.
Select **Design Options** on the Design Options panel.

93. 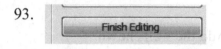 Select **Finish Editing**.
Close the dialog.

94. Verify the Design Option is now set to **Main Model**.

95. 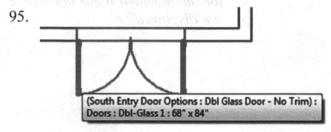 Note that if you hover your mouse over the element, it will display which Option set it belongs to.

This only works if Active Only or Exclude Options is disabled.

96. Activate the Sheet named **Office Layout Options**.

97. Drag and drop the two Office Layout options onto the sheet.

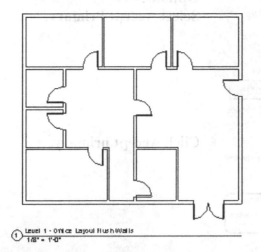 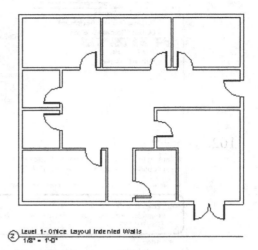

98. Activate the Manage tab on tab on ribbon.
Select **Design Options**.

99. *Let's assume that the client decided they prefer the flush walls option.*

Highlight the **Flush Walls** option.

100. Select **Make Primary**.

Note that (primary) is now next to Flush Walls.

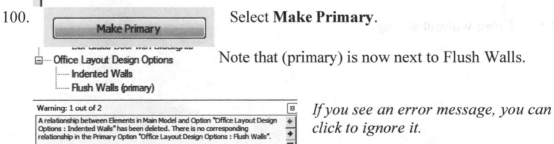

If you see an error message, you can click to ignore it.

101.

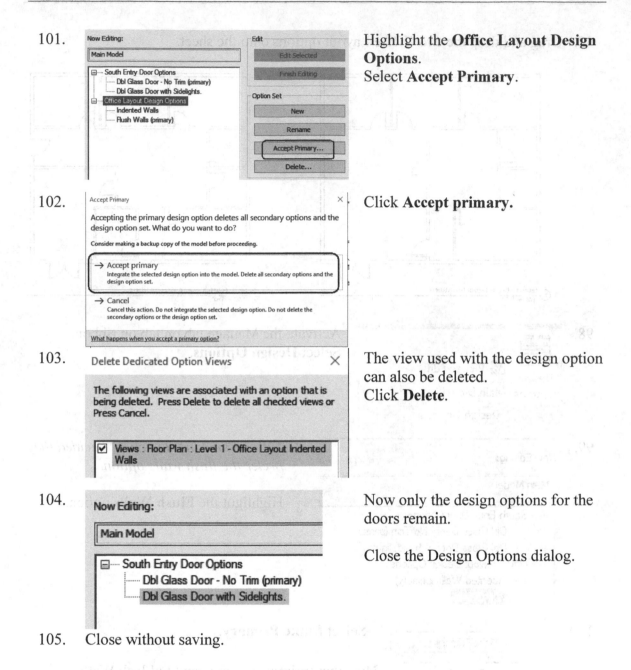

Highlight the **Office Layout Design Options**.
Select **Accept Primary**.

102.

Click **Accept primary.**

103.

The view used with the design option can also be deleted.
Click **Delete**.

104.

Now only the design options for the doors remain.

Close the Design Options dialog.

105.   Close without saving.

Exercise 5-12

Design Options Practice Question

Drawing Name: i_Design_Options_Question.rvt
Estimated Time to Completion: 5 Minutes

Scope
Design Options
Properties

Solution

1. 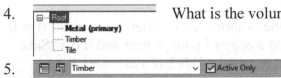 Select the **Open** tool.

2. File name: | i_Design_Options_Question.rvt | Locate the *i_Design_Options_Question* file.
Select Open.

3.
- 3D Views
  - Approach
  - From Yard
  - Kitchen
  - Living Room
  - Living Room - ISO
  - Section Perspective
  - Solar Analysis
  - {3D}

Activate the {3D} view under 3D Views.

4.
- Roof
  - **Metal (primary)**
  - Timber
  - Tile

What is the volume of the roof in Design Option – Timber?

5. Timber ☑ Active Only   Select the **Timber** design option.
Enable **Active Only**.

6.

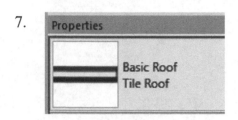

Select the roof.
Go to the Properties panel.

Scroll down.

What is the volume of the roof?

7.

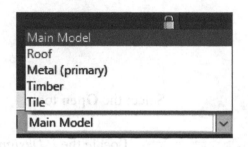

If you switch design options, does the volume of the roof change?

Phases

Phases are distinct, separate time periods within the life of a building. Phases can represent either the time periods or how the building appeared during that time period. By default, every Revit project has two phases or time periods already pre-defined. They are named Existing and New Construction.

The most common use of phases is to keep track of the "Before" and "After" scenarios. If you are a recovering AutoCAD user, you probably created a copy of your project and did a "Save As" to re-work the existing building for the proposed remodel. This has a number of disadvantages – not the least of which is the use of external references, the possibility of missing something, and the duplication of data.

If you are working on a completely new building project on a "clean" site, you still might want to use Phases as a way to control when to schedule special equipment on the site as well as crew. You might have a phase for foundation work, a phase for framing, a phase for electrical and so on. You can create schedules based on phases, so you will know exactly what inventory you might need on hand based on the phase.

You determine what elements are displayed in a view by assigning a phase to the view. You can even control colors and linetypes of different phases so you get a visual cue on which elements were created or placed in which phase.

EXISTING PLAN

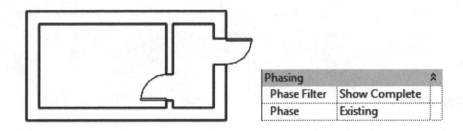

DEMOLITION PLAN

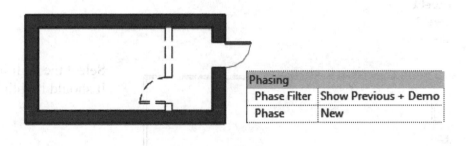

NEW FLOOR PLAN

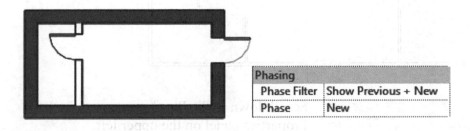

A common error is to create a phase for demolition. This is unnecessary. Instead you can duplicate the view and set one view as the Demolition Plan and one view at the new floor plan.

Exercise 5-13
Phases

Drawing Name: c_phasing.rvt
Estimated Time to Completion: 75 Minutes

Scope
Properties
Filter
Phases
Rename View
Copy View
Graphic Settings for Phases

Solution

1. Activate **Level 1** under Floor Plans.

    Views (all)
      Floor Plans
        Level 1
        Level 2
        Site

2. 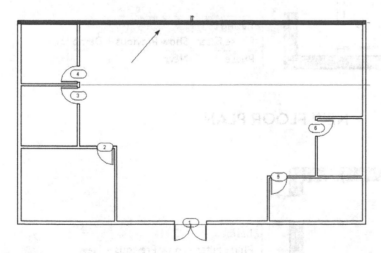 Select the wall indicated. It should highlight.

3. Scroll down to the Phasing category in the Properties panel on the upper left.

    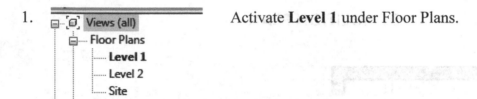

    This wall was created in the New Construction Phase.
    Note that it is not set to be demolished.

4. Right click and Click **Cancel** to deselect the wall.

5. 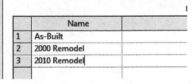 Go to the **Manage** tab on tab on ribbon.
Select **Phasing→Phases**.

6. Rename Existing to **As-Built**.

	Name
1	As-Built
2	New Construction

7. Rename New Construction to **2000 Remodel**.

	Name
1	As-Built
2	2000 Remodel

8.  Highlight the **2000 Remodel**.
Select **After**.

9. Name the new phase **2010 Remodel**.

	Name
1	As-Built
2	2000 Remodel
3	2010 Remodel

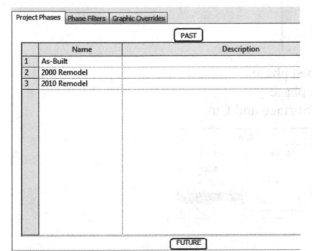

Note that the top indicates the past and the bottom indicates the future to help orient the phases.

10. Select the **Graphic Overrides** tab.

Phase Status	Projection/Surface		Cut		Halftone	Material
	Lines	Patterns	Lines	Patterns		
Existing	———————		———————	Hidden	☐	Phase-Exist
Demolished	- - - - - - - -		- - - - - - - -	Hidden	☐	Phase-Demo
New	———————		———————	▮▮▮▮	☐	Phase-New
Temporary	·············		····················	/////	☐	Phase-Temp

11. Note that in the Lines column for the Existing Phase, the line color is set to gray.

12.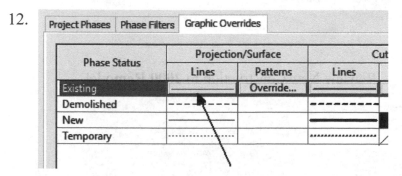

Highlight **Existing**. Click in the **Lines** column and the Line Graphics dialog will display.

Projection/Surface is what is displayed in the floor plan views.
Cut is the display for elevation or section views.
Override indicates you have changed the display from the default settings.

13. Set the Color to **Green** for the Existing phase by selecting the color button. Click **OK**.

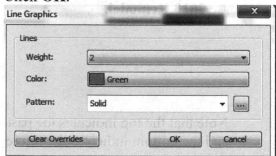

Set the Color to **Blue** for the Demolished phase.
Set the Color to **Magenta** for the New phase.
Change the colors for both Projection/Surface and Cut.

Phase Status	Projection/Surface		Cut	
	Lines	Patterns	Lines	Patterns
Existing	———————		———————	Hidden
Demolished	- - - - - - - -		- - - - - - - -	Hidden
New	———————	Override...	———————	▮▮▮▮
Temporary	·············		····················	/////

14. **Project Phases** **Phase Filters** **Graphic Overrides**

Select the **Phase Filters** tab.

15.

	Filter Name	New	Existing	Demolished	Temporary
1	Show All	By Category	Overridden	Overridden	Overridden
2	Show Demo + New	By Category	Not Displayed	Overridden	Overridden
3	Show Previous + Dem	Not Displayed	Overridden	Overridden	Not Displayed
4	Show Previous + New	By Category	Overridden	Not Displayed	Not Displayed
5	Show Previous Phase	Not Displayed	Overridden	Not Displayed	Not Displayed

Note that there are already phase filters pre-defined that will control what is displayed in a view.

16. **New**   Click the **New** button on the bottom of the dialog.

17.

	Filter Name
1	Show All
2	Show Demo + New
3	Show Previous + Dem
4	Show Previous + New
5	Show Previous Phase
6	Show Existing

Change the name for the new phase filter to **Show Existing**.
Show Previous + Demo will display existing plus demo elements, but not new.
Show Previous + New will display existing plus new elements, but not demolished elements.

18. In the New column, select **Overridden**.
In the Existing column, select **Overridden**.
This means that the default display settings will use the new color assigned.
In the Demolished column, select **Not Displayed**.

	Filter Name	New	Existing	Demolished
1	Show All	By Category	Overridden	Overridden
2	Show Demo + New	By Category	Not Displayed	Overridden
3	Show Previous + Dem	Not Displayed	Overridden	Overridden
4	Show Previous + New	By Category	Overridden	Not Displayed
5	Show Previous Phase	Not Displayed	Overridden	Not Displayed
6	Show Existing	Overridden	Overridden	Not Displayed

Phasing

	Filter Name	New	Existing	Demolished	Temporary
1	Show All	By Category	Overridden	Overridden	Overridden
2	Show Demo + New	Overridden	Overridden	Overridden	Overridden
3	Show Existing	Not Displayed	Overridden	Not Displayed	Overridden
4	Show Previous + Demo	Not Displayed	Overridden	Overridden	Not Displayed
5	Show Previous + New	Overridden	Overridden	Not Displayed	Not Displayed
6	Show Previous Phase	Not Displayed	Overridden	Not Displayed	Not Displayed

19. Use Overridden to display the colors you assigned to the different phases.
Verify that in the Show Previous + Demo phase New elements are not displayed.
Verify that in the Show Previous + New phase Demolished elements are not displayed.
Click **Apply** and **OK** to close the Phases dialog.

20.
Window around the entire floor plan.
Select the **Filter** button.

21.
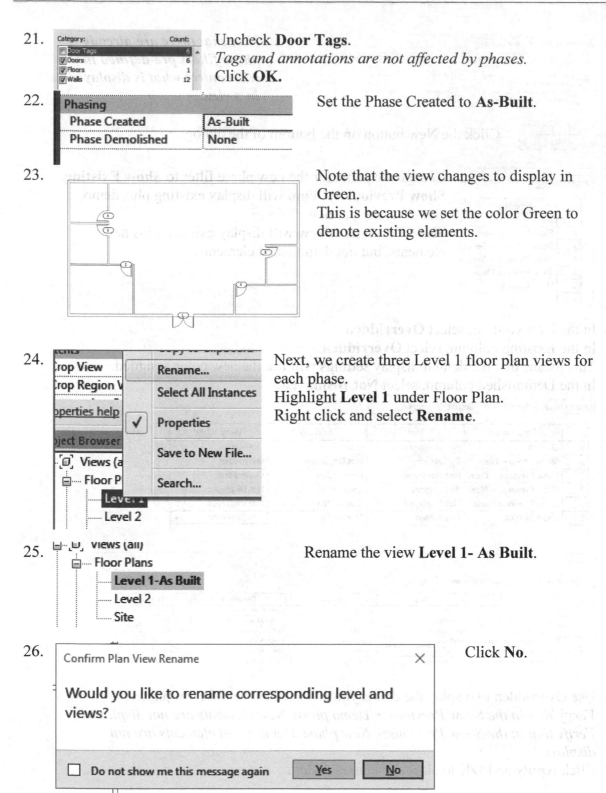

Uncheck **Door Tags**.
Tags and annotations are not affected by phases.
Click **OK.**

22.

Set the Phase Created to **As-Built**.

Phasing	
Phase Created	As-Built
Phase Demolished	None

23.

Note that the view changes to display in Green.
This is because we set the color Green to denote existing elements.

24.

Next, we create three Level 1 floor plan views for each phase.
Highlight **Level 1** under Floor Plan.
Right click and select **Rename**.

Rename...
Select All Instances
✓ Properties
Save to New File...
Search...

Crop View
Crop Region V
operties help
ject Browser
Views (a
Floor P
Level 1
Level 2

25.

Rename the view **Level 1- As Built.**

views (all)
Floor Plans
Level 1-As Built
Level 2
Site

26.

Click **No.**

Confirm Plan View Rename   ✕

Would you like to rename corresponding level and views?

☐ Do not show me this message again   Yes   No

27. 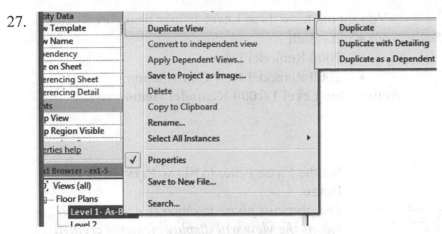 Highlight **Level 1-As Built** under Floor Plan.
Right click and select **Duplicate View→Duplicate**.

28. 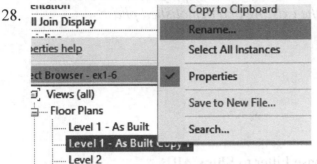 Highlight **Level 1-As Built Copy 1** under Floor Plan.
Right click and select **Rename**.

29. 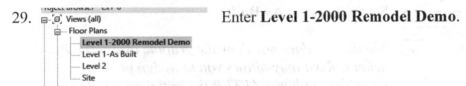 Enter **Level 1-2000 Remodel Demo**.

30. Highlight **Level 1-As Built** under Floor Plan.

31. Right click and select **Duplicate View→Duplicate**.

32. 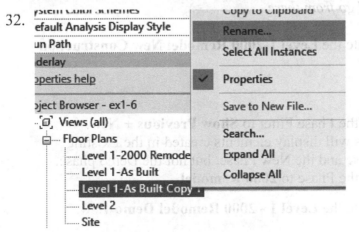 Highlight **Level 1-As Built Copy 1** under Floor Plan.
Right click and select **Rename**.

33. Enter **Level 1-2000 Remodel New Construction**.

34. You should have three Level 1 floor plan views listed:
  - As Built
  - 2000 Remodel Demo
  - 2000 Remodel New Construction.

  Activate the **Level 1-2000 Remodel Demo** view.

35.

36. In the Properties dialog:

  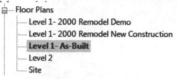

  Set the Phase to **2000 Remodel**.

  Set the Phase Filter to **Show Previous + Demo**.
  The previous phase to demo is As-Built. This means the view will display elements created in the existing and demolished phase.

37. Activate the **Level 1-As Built** view.

38. In the Properties dialog:

Phasing	
Phase Filter	Show All
Phase	As-Built

  Set the Phase Filter to **Show All**.
  Set the Phase to **As-Built**.

  The display does not show the graphic overrides. By default, Revit only allows you to assign graphic overrides to phases AFTER the initial phase. Because the As-Built view is the first phase in the process, no graphic overrides are allowed. The only work-around is to create an initial phase with no graphic overrides and go from there.

39. Activate the **Level 1-2000 Remodel New Construction** view.

40. In the Properties dialog:

  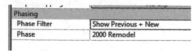

  Set the Phase Filter to **Show Previous + New**.
  This will display elements created in the Existing Phase and the New Phase, but not the Demo phase.
  Set the Phase to **2000 Remodel**.

41. Activate the **Level 1 - 2000 Remodel Demo** view.

42. 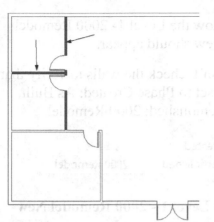 Hold down the Ctrl button.
    Select the two walls indicated.

43. In the Properties pane: Scroll down to the bottom.
    In the Phase Demolished drop-down list, select **2000 Remodel**.

    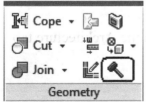

44. The demolished walls change appearance based on the graphic overrides.
    Release the selected walls using right click→Cancel or by Clicking ESCAPE.

    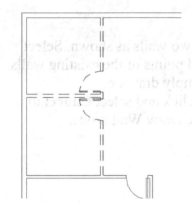

45. 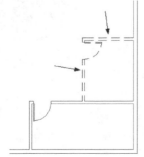 Activate the **Modify** tab on tab on ribbon.
    Use the **Demolish** tool on the Geometry panel to demolish the walls indicated.

46. Note that the doors will automatically be demolished along with the walls. If there were windows placed, these would also be demolished. That is because those elements are considered *wall-hosted*.

    Right click and select Cancel to exit the Demolish mode.

47.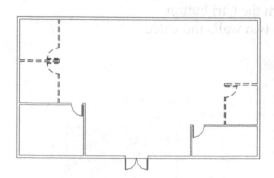

This is how the Level 1- 2000 Remodel Demo view should appear.

If it doesn't, check the walls to verify that they are set to Phase Created: As Built, Phase Demolished: 2000 Remodel.

Phasing	
Phase Created	As-Built
Phase Demolished	2000 Remodel

48.
- Floor Plans
  - Level 1- 2000 Remodel Demo
  - **Level 1- 2000 Remodel New Construction**
  - Level 1- As-Built
  - Level 2

Activate the **Level 1 - 2000 Remodel New Construction** view.

49.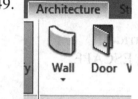

Select the **Wall** tool from the Architecture tab on tab on ribbon.

50.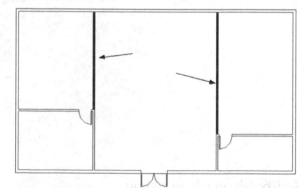

Place two walls as shown. Select the end points of the existing walls and simply draw up.
Right click and select **Cancel** to exit the Draw Wall mode.

51. Architecture St

Wall Door

Select the **Door** tool under the Build panel on the Architecture tab on tab on ribbon.

52.

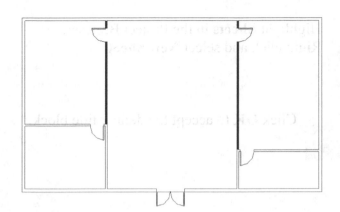

Place two doors as shown. Set the doors 3′ 6″ from the top horizontal wall. Flip the orientation of the doors if needed.

You can Click the space bar to orient the doors before you left click to place.

Note that the new doors and walls are a different color than the existing walls.

53.

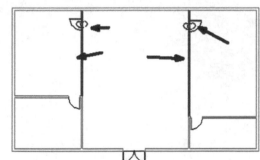

Select the doors and walls you just placed. You can select by holding down the CONTROL key or by windowing around the area.

Note: If Door Tags are selected, you will not be able to access Phases in the Properties dialog.

54. Look in the Properties panel and scroll down to Phasing.

Phasing	
Phase Created	2000 Remodel
Phase Demolished	None

Note that the elements are already set to **2000 Remodel** in the Phase Created field.

55. Switch between the three views to see how they display differently.

56. Remember that Existing should show as Green, Demo as Blue, and New as Magenta.

If the colors don't display correctly in the Level 1 Remodel Demo or Remodel New Construction, check the Phase Filters again and make sure that the categories to be displayed are Overridden to use the assigned colors.

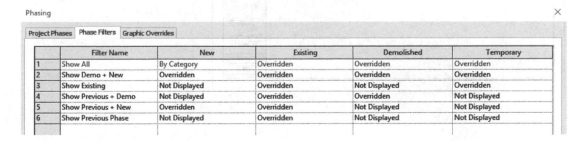

Phasing ×

Project Phases | Phase Filters | Graphic Overrides

	Filter Name	New	Existing	Demolished	Temporary
1	Show All	By Category	Overridden	Overridden	Overridden
2	Show Demo + New	Overridden	Overridden	Overridden	Overridden
3	Show Existing	Not Displayed	Overridden	Not Displayed	Overridden
4	Show Previous + Demo	Not Displayed	Overridden	Overridden	Not Displayed
5	Show Previous + New	Overridden	Overridden	Not Displayed	Not Displayed
6	Show Previous Phase	Not Displayed	Overridden	Not Displayed	Not Displayed

57.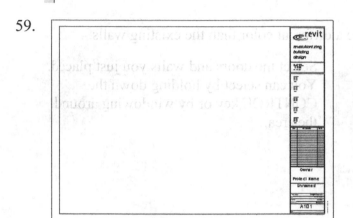

Highlight **Sheets** in the Project Browser.
Right click and select **New Sheet**.

58.

Click **OK** to accept the default title block.

59.

A view opens with the new sheet.

60.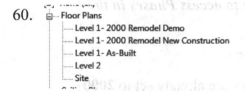

Highlight the Level 1 – As-Built Floor plan.
Hold down the left mouse button and drag the view onto the sheet. Release the left mouse button to click to place.

61. A preview will appear on your cursor. Left click to place the view on the sheet.

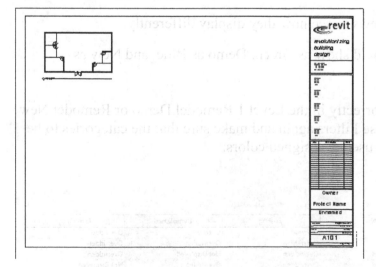

62. Highlight the Level 1 - 2000 Remodel Demo Floor plan.
    Hold down the left mouse button and drag the view onto the sheet. Release the left mouse button to click to place.

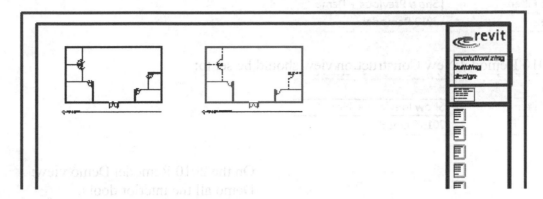

The two views appear on the sheet.

63. Highlight the Level 1 - 2000 New Construction plan.
    Hold down the left mouse button and drag the view onto the sheet. Release the left mouse button to click to place.

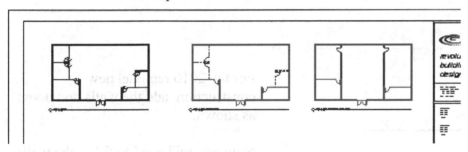

64.   Save as *ex5-13.rvt*.

Challenge Exercise:

Create two more views called Level 1 2010 Remodel Demo and Level 1 2010 Remodel New Construction.

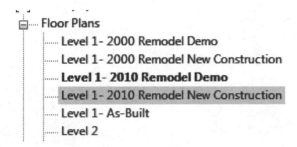

Set the Phases and phase filters to the new views.

The 2010 Remodel Demo view should be set to:

Phasing	
Phase Filter	Show Previous + Demo
Phase	2010 Remodel

The 2010 Remodel New Construction view should be set to:

Phasing	
Phase Filter	Show Previous + New
Phase	2010 Remodel

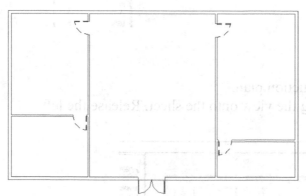

On the 2010 Remodel Demo view:
Demo all the interior doors.

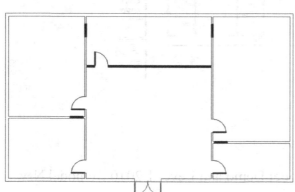

For the 2010 remodel new construction, add the walls and doors as shown.

Note you will need to fill in the walls where the doors used to be.

Add the 2010 views to your sheet.

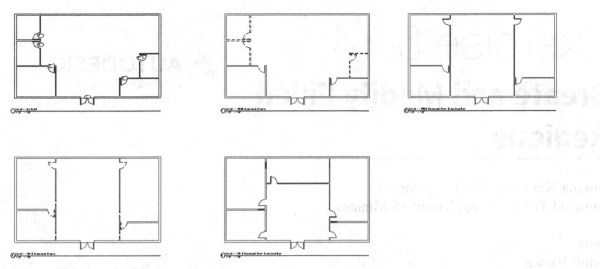

Answer this question:

When should you use phasing as opposed to design options?

Filled Regions

The Filled Region tool creates a 2-dimensional, view-specific graphic with a boundary line style and fill pattern within the closed boundary. The tool is useful for defining a filled area in a detail view or for adding a filled region to an annotation family.

The filled region is parallel to the view's sketch plane.

A filled region contains a fill pattern. Fill patterns are of 2 types: Drafting or Model. Drafting fill patterns are based on the scale of the view. Model fill patterns are based on the actual dimensions in the building model.

A filled region created for a detail view is part of the Detail Items category. Revit lists the region in the Project Browser under Families ➤ Detail Items ➤ Filled Region. If you create a filled region as part of an annotation family, Revit identifies it as a Filled Region, but does not store it in the Project Browser.

Exercise 5-14
Create and Modify Filled Regions

Drawing Name: m_filled_regions.rvt
Estimated Time to Completion: 15 Minutes

Scope
Detail Views
Filled Regions

Solution

1. Go to the View tab on the ribbon.

   Select **Drafting View**.

2. Type **Parapet Detail 1** in the Name field.

   Click **OK**.

Name:	Parapet Detail 1
Scale:	1 : 10
Scale value 1:	10

3. Views (all)
   - Floor Plans
     - Level 1
     - Level 2
     - Site
   - Ceiling Plans
     - Level 1
     - Level 2
   - Elevations (Building Elevation)
     - East
     - North
     - South
     - West
   - Drafting Views (Detail)
     - **Parapet Detail 1**

   Notice the new drafting view is listed in the Project Browser.

4. Switch to the Insert tab on the ribbon.

   Select **Import CAD**.

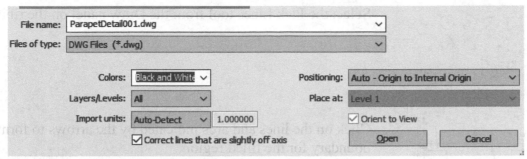

5. Locate the *ParapetDetail001.dwg* file.
    Set the Colors to **Black and White**.
    Set Layers to **All**.
    Set Import Units to **Auto-Detect**.
    Set Positioning to **Auto- Origin to Internal Origin**.
    Click **Open**.

6. 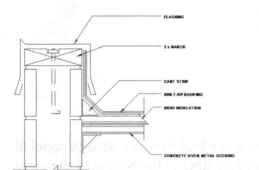 Double click the wheel on your mouse to Zoom to Fit.

7. 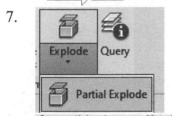 Select the imported CAD object.

    Click **Partial Explode** on the ribbon.

8. 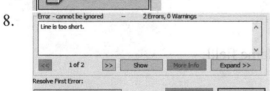 An error dialog appears.

    Click **Delete Elements**.

9. 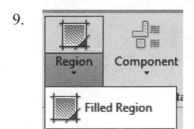 Switch to the Annotate tab on the ribbon.
    Select **Filled Region**.

10. 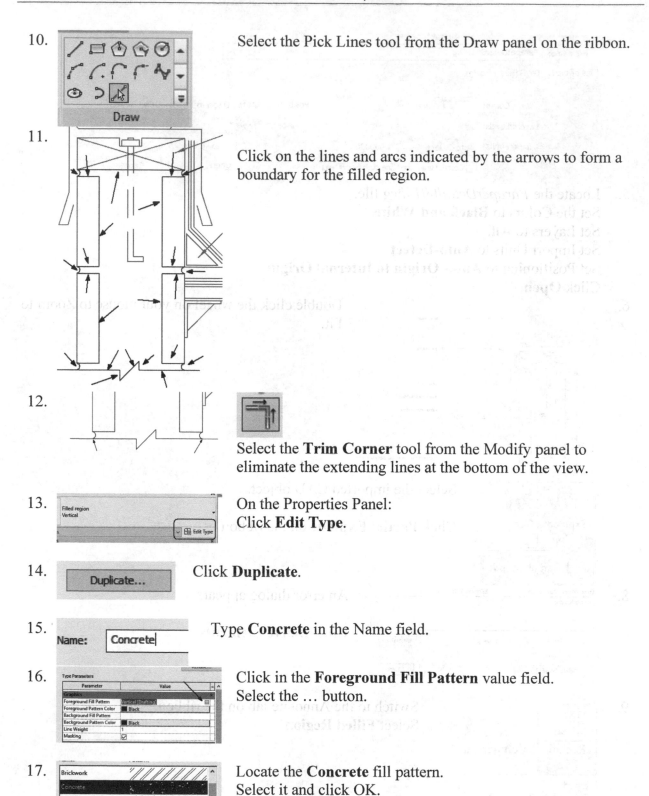 Select the Pick Lines tool from the Draw panel on the ribbon.

11. Click on the lines and arcs indicated by the arrows to form a boundary for the filled region.

12. Select the **Trim Corner** tool from the Modify panel to eliminate the extending lines at the bottom of the view.

13. On the Properties Panel:
Click **Edit Type**.

14. Click **Duplicate**.

15. Type **Concrete** in the Name field.

Name: Concrete

16. Click in the **Foreground Fill Pattern** value field.
Select the ... button.

17. Locate the **Concrete** fill pattern.
Select it and click OK.

18.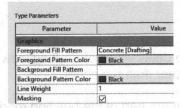

Verify that the correct fill pattern is displayed in the Type Parameters dialog.

Click **OK**.

19.

The Properties Panel should display the desired fill region.

Click the **Green Check** on the ribbon.

20.

If you see this error, it means you picked the same line more than once.

Click **Continue** and delete the overlapping lines.

The overlapping lines are highlighted.

21.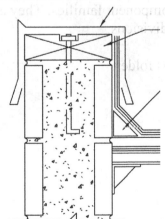

The filled region is created.

Save as *ex5-14.rvt*.

Detail Components

Detail components are line-based 2D elements that you can add to detail views or drafting views. They are visible only in those views. They scale with the model, rather than the sheet.

Detail components are not associated with the model elements that are part of the building model. Instead, they provide construction details or other information in a specific view.

For example, in the following drafting view, the studs, insulation, and siding are detail components.

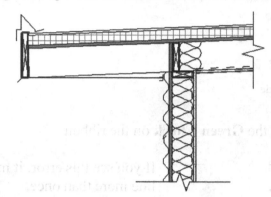

You can tag detail components with detail item tags, or you can keynote detail components.

Before adding detail components to a view, load the desired detail component families into the project from the family library. Revit contains over 500 detail component families. They are organized by the 16 CSI (Construction Specifications Institute) divisions.

Revit's detail component library is located under the Detail Items folder.

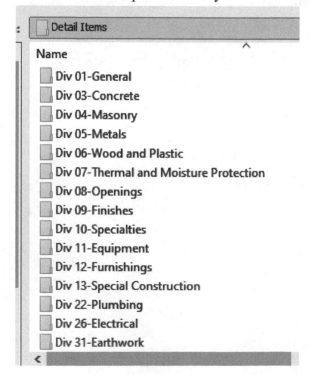

Exercise 5-15

Detail Components

AUTODESK.
Certified Professional

Drawing Name: **i_detail_components.rvt**
Estimated Time to Completion: 45 Minutes

Scope

Add detail components to a detail view.
Set a detail view to independent.
Change the clip offset of a detail view.
Use Repeating Details.
Use detail lines.

Solution

1. Views (all)
    Floor Plans
    —— Ground Floor
    —— Lower Roof
    —— Main Floor
    —— **Main Roof**
    —— Site
    —— T. O. Footing
    —— T. O. Parapet

    Activate the **Main Roof** floor plan.

2.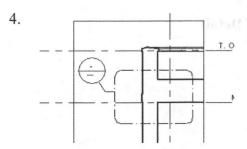

    Double click on the section head located between A-B grids to activate that section view.

3. —— Callout of Section 2
    Sections (Wall Section)
    —— **Section 2**
    —— Section 4

    Note the name of the active section view in the browser.

4.

    Double click on the callout bubble to activate the callout view.

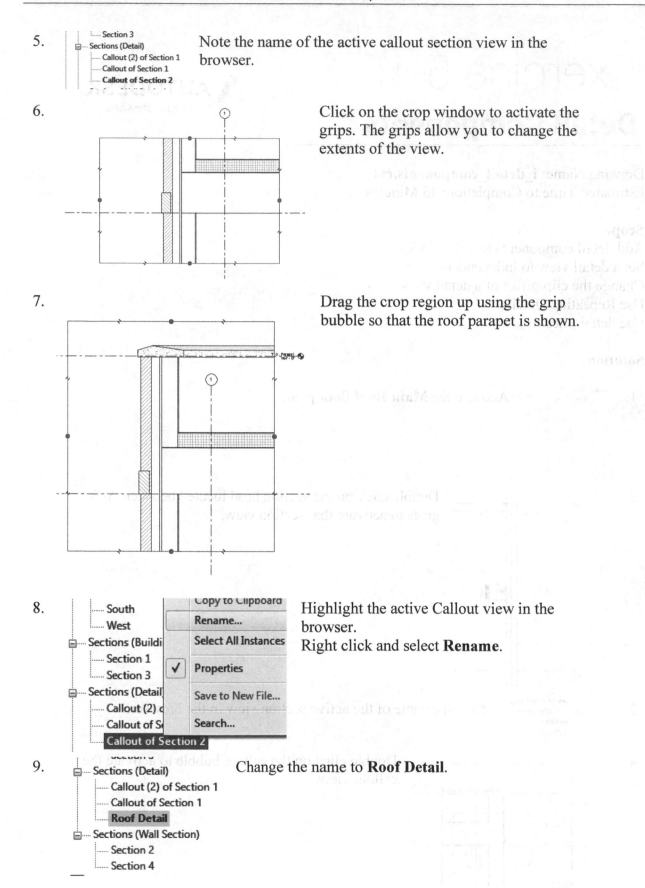

5. Note the name of the active callout section view in the browser.

6. Click on the crop window to activate the grips. The grips allow you to change the extents of the view.

7. Drag the crop region up using the grip bubble so that the roof parapet is shown.

8. Highlight the active Callout view in the browser.
    Right click and select **Rename**.

9. Change the name to **Roof Detail**.

10.
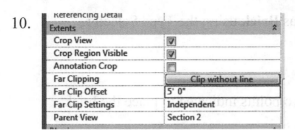

In the Properties pane,
under Extents,
set the Far Clip Settings to **Independent**.
Change the Far Clip Offset to **5' 0"**.

Note that the view updates and no longer
displays the far North wall.

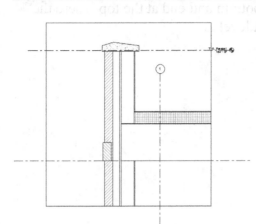

11.
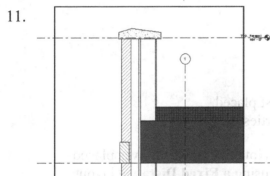

Select the roof.

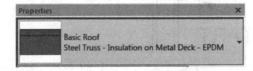

If the roof is selected, it should be listed in
the Properties pane.

12.
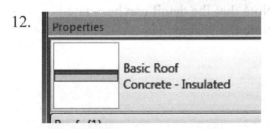

Change the Roof type to **Concrete - Insulated**
on the Properties pane.

*If the roof is changed in the section view, is it
changed in the model?*

13.
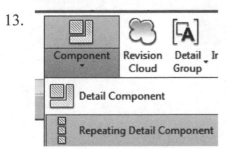

Select the **Repeating Detail Component** tool from
the Detail panel on the Annotate tab on ribbon.

*A repeating detail component is an array spaced
equally using a detail component family. Any detail
component family can be used.*

14. **Repeating Detail Brick**   Set the repeating detail as **Brick** using the Type Selector.

15. 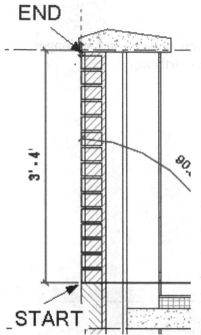   Pick the two points indicated to place the brick detail.
     Start at the bottom and end at the top where the T.O. Parapet level is.

16. Edit Type   Select the brick repeating detail just placed.
     Select the **Edit Type** on the Properties pane.

17.

Parameter	
Pattern	
Detail	Brick Standard : Running Section
Layout	Fixed Distance
Inside	☑
Spacing	0' 2 5/8"
Detail Rotation	None

Note that the detail was placed using a **Fixed Distance** layout.
Note that the detail can be rotated.
Note the spacing is set to **2 5/8"**.

18. Duplicate...   Select **Duplicate**.

19. Name: Mortar   Type **Mortar**.
     Click **OK**.

20.

Pattern	
Detail	Mortar Joint : Brick Joint
Layout	Fixed Distance
Inside	☑
Spacing	0' 2 5/8"
Detail Rotation	None

For the Detail field, select **Mortar Joint: Brick Joint**. Click **OK**.
If you accidently change the brick repeating detail to the new repeating detail, change it back using the type selector.

21. Component

Select the **Repeating Detail** tool from the Detail panel on the Annotate tab on ribbon.

22. Repeating Detail Mortar

Set the repeating detail as **Mortar** using the Type Selector.

23. Start on top of the brick placed and end above the Main Roof level.

24. Zoom in to see the mortar joints.

25. Component

Select the **Detail Component** tool from the Annotate tab on ribbon.

26. Cant Strip 3" x 3"

Select the **3" x 3" Cant Strip** from the drop-down list on the Type Selector.

27. Place the cant strip as shown.

28. Select the **Detail Line** tool from the Annotate tab on ribbon.

    Detail
    Line

29. Select **Medium Lines** from the Line Style options drop-down list.

    Line Style:
    Medium Lines

    Line Style

30. Enable **Chain**. Enter **1/4″** in the Offset box.

    ☑ Chain Offset: 0' 0 1/4"

31. Select **Pick Lines** mode from the Draw panel.

    Draw

32. 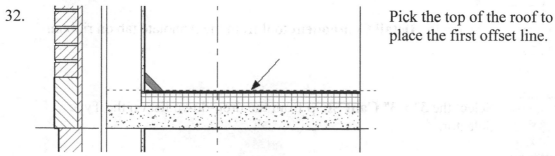 Pick the top of the roof to place the first offset line.

33. 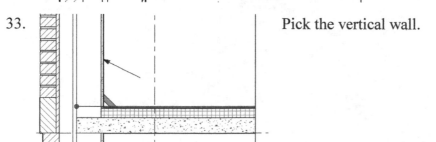 Pick the vertical wall.

34. 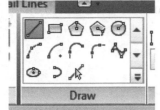 Switch to **Draw Line** mode.

35. 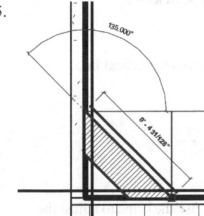 Place a line parallel to the Cant Strip.
Set the offset to 1/4" and start the line at the top of the cant strip. This will orient the line above the cant strip.

36. 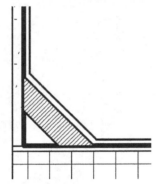 Use the **Trim** tool on the Modify Tab on ribbon to clean up the detail lines.

37. 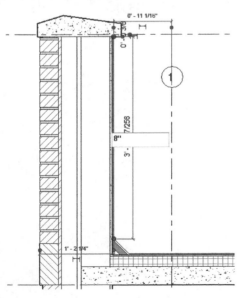 Select the vertical line to enable the temporary dimension.
Select the vertical line's vertical dimension.
Change the value to 8".

38. 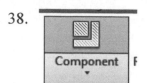 Select **Component→Detail Component** on the Annotate tab on ribbon.

39. 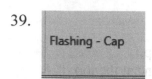 In the Properties pane,
select **Flashing - Cap** using the Type Selector.

40. 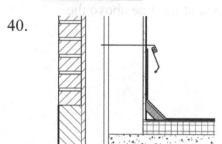 Place the Flashing Cap above the vertical line.

41. 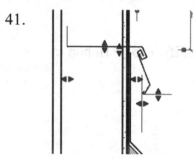 Select the flashing cap and use the grips to adjust the size.

42. Go to **File→Save As**. Save the file as *ex5-15.rvt*.

Exercise 5-16

Tag Elements by Category

AUTODESK.
Certified Professional

Drawing Name: tag_elements.rvt
Estimated Time: 5 minutes

This exercise reinforces the following skills:

- ❑ Tag All Not Tagged
- ❑ Window Schedules
- ❑ Schedule/Quantities
- ❑ Schedule Properties

1.

{3D}
Elevations (Building Elevation)
├── East
├── **North**
├── South
├── South Lobby
└── West
Activate the **North Building Elevation**.

2.
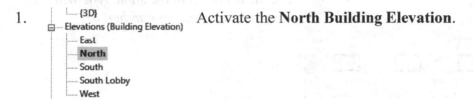
Tag by Category
Activate the Annotate tab on ribbon.

Select the **Tag by Category** tool from the Tag panel.

3. `☑ | Tags... | ☐ Leader | Attached End | ⌄ | ⊢`   Disable **Leader** on the Options bar.

4. Click on a window in the view to place the window tag. Right click and select Cancel to exit the command.

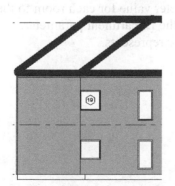

Revit automatically knew to place a window tag based on what was selected.

5. Activate the Annotate tab on ribbon.

Select the **Tag All** tool from the Tag panel.

6. 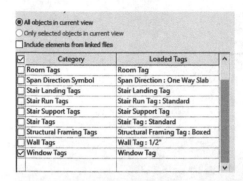 Click **Window Tags**.

Click **OK**.

You can select more than one category tag to be placed using Tag All.

7. 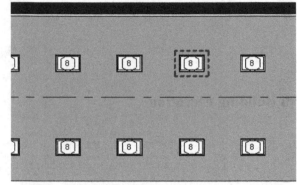 Tags will appear on all the windows.

The window tag uses the window type as an identifier, not the window mark. So, all windows of the same type will display the same tag number.

8. Save the file as *ex5-16.rvt*.

Color Schemes

Colors schemes are useful for graphically illustrating categories of spaces, rooms or areas.

For example, you can create a color scheme by room name, area, occupancy, or department. If you want to color rooms in a floor plan by department, set the Department parameter value for each room to the necessary value, and then create a color scheme based on the values of the Department parameter. You can then add a color fill legend to identify the department that each color represents.

Color schemes can be applied to:

- Rooms
- Areas
- Spaces or Zones
- Pipes or ducts

Exercise 5-17

Color Schemes

AUTODESK.
Certified Professional

Drawing Name: **i_Color_Scheme.rvt**
Estimated Time to Completion: 15 Minutes

Scope
Create an area plan using a color scheme.
Place a color scheme legend.

Solution

1. Floor Plans
    01 - Entry Level
    01 - Entry Level-Rooms
    02 - Floor
    03 - Floor
    Roof

    Activate the **01- Entry Level** floor plan.

2. | Show Hatch Lines | By Discipline |
    | Color Scheme Location | Background |
    | Color Scheme | <none> |
    | System Color Schemes | Edit... |
    | Default Analysis Display Style | None |

    In the Properties pane, scroll down to the **Color Scheme** field.
    Click **<none>**.

3. Schemes
    Category:
    Rooms

    (none)
    Name
    Department

    Select the **Rooms** category.
    Highlight **Name**.

4. Scheme Definition
    Title: Color:
    Area Legend   Area

    In the Title field, enter **Area Legend**.
    Under Color, select the **Area** scheme.

5. Colors are not preserved when changing which parameter is colored. To color by a different parameter, consider making a new color scheme.

    Click **OK**.

6.

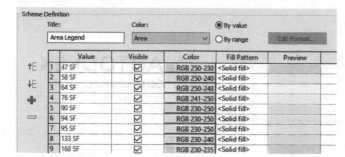

Note that different colors are applied depending on the square footage of the rooms.

Click **OK**.

7.

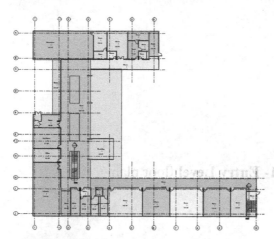

The rooms fill in according to the square footage.

8.

Activate the **Annotate** tab on ribbon.
Select the **Color Fill Legend** tool on the Color Fill panel.

9.

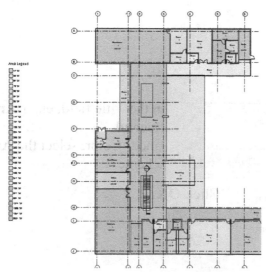

Place the legend in the view.

10.  Select the legend that was placed in the display window.
Select **Edit Scheme** from the Scheme panel on the tab on ribbon.

11. Change the Fill Pattern for 47 SF **to Crosshatch-Small**.

	Value	Visible	Color	Fill Pattern	Preview
1	47 SF	☑	■ RGB 156-185	Crosshatch-small	
2	58 SF	☑	■ PANTONE 3	Solid fill	
3	64 SF	☑	■ PANTONE 6	Solid fill	

12. Change the hatch patterns for the next few rows. Click **OK**.

	Value	Visible	Color	Fill Pattern	Preview
1	47 SF	☑	■ RGB 156-185	Crosshatch-small	
2	58 SF	☑	■ PANTONE 3	Crosshatch	
3	64 SF	☑	■ PANTONE 6	Diagonal crosshatch	
4	76 SF	☑	■ RGB 139-166	Diagonal down	
5	90 SF	☑	■ PANTONE 6	Solid fill	
6	94 SF	☑	■ RGB 096-175	Solid fill	

13. *Note that the legend updates.*

Area Legend

☐ 47 SF

☐ 58 SF

☐ 64 SF

☐ 76 SF

14.  Select the Color Fill Legend. On the Properties pane, select **Edit Type.**

15. 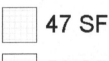 Change the font for the Text to **Tahoma.**
Enable **Bold**.

16.

Title Text	
Font	Tahoma
Size	1/4"
Bold	☑
Italic	☑
Underline	☐

Change the font for the Title Text to **Tahoma.**
Enable **Bold.**
Enable **Italic.**
Click **OK**.

17.

Area Legend
☐ 47 SF
☐ 58 SF
☐ 64 SF

Note how the legend updates.

18. Close without saving.

Certified User Practice Exam

1. When you add text notes to a view using the Text tool, you can control all of the following except:

    A. The display of leader lines
    B. Text Wrapping
    C. Text Font
    D. Whether the text is bold or underlined

2. To use a different text font, you need to:

    A. Use the Text tool.
    B. Create a new Text family and change the Type Properties.
    C. Create a new Text family and change the Instance Properties.
    D. Modify the text.

3. _____ are the measurements that display in the drawing when you select an element. These dimensions disappear when you complete the action or deselect the element.

    A. Temporary
    B. Permanent
    C. Listening
    D. Constraints

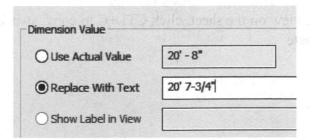

4. You click on a permanent dimension to change the value. You enable Replace With Text and type in the desired value. Then you click OK. What happens next?

    A. An error message appears advising that you cannot modify a dimension value using text
    B. The dimension updates with the new value you entered
    C. The dimension updates and the elements shift location
    D. The dimension updates, but the element remains in the same position

5. To add a new sheet to a project: (select all that apply)

    A. Right click on the Sheets category in the Project Browser and select New Sheet
    B. Select the New Sheet tool on the View tab on the ribbon
    C. Select New Sheet tool on the Annotation tab on the ribbon
    D. Select New Sheet tool on the Manage tab on the ribbon

6. To change the size of a sheet used in a Revit project:

    A. Right click on the sheet in the Project Browser and select Properties
    B. Highlight the sheet in the Project Browser and select Edit Type in the Properties panel
    C. Select the title block on the sheet, select a new title block using the Type Selector
    D. Highlight the sheet in the Project Browser and use the Type Selector to change the size

7. To add a plan view to a sheet: (select all that apply)

    A. Drag and drop a view from the Project Browser on to a sheet
    B. Select the Place View tool from the View tab on the ribbon
    C. Select the Insert View tool from the Insert tab on the ribbon
    D. Select the Place View tool from the Insert tab on the ribbon

8. You want to place the Level 1 floor plan view on more than one sheet. The best method to use is:

    A. Duplicate the view so it can be placed on more than one sheet
    B. Drag the view on each desired sheet
    C. Highlight the view on the sheet, select Edit Type on the Properties Panel, select Duplicate
    D. Highlight the view on the sheet, click CTL+C to copy, switch to the other sheet, click CTL+V to paste

ANSWERS: 1) D; 2) B; 3) A; 4) A; 5) A & B; 6) C; 7) A & B; 8) A

Certified Professional Practice Exam

1. Color Schemes can be placed in the following views (Select 3):

    A. Floor plan
    B. Ceiling Plan
    C. Section
    D. Elevation
    E. 3D

2. To define the colors and fill patterns used in a color scheme legend, select the legend, and Click Edit Scheme on the:

    A. Properties palette
    B. View Control Bar
    C. Ribbon
    D. Options Bar

3. A legend is a view that can be placed on:

    A. A plan view
    B. A drafting view
    C. Multiple Plan Regions
    D. Multiple Sheets

4. To change the font used in a Color Scheme Legend:

    A. Select the legend, then select Edit Scheme from the ribbon.
    B. Select the text in the legend, double click to edit.
    C. Select the legend, then select Edit Type from the Properties pane.
    D. Select the legend, right click and select Edit Family.

5. Filled regions can be placed in the following views: (Select all that apply)

    A. Floor Plan
    B. Elevation
    C. Section
    D. Detail View

6. Filled regions: (select all that apply)

    A. Are two-dimensional elements
    B. View-specific (only visible in the view where they are created and placed)
    C. Use fill patterns
    D. Can be placed in 3D sections

7. Detail Components are:

   A. System Families
   B. Mass Families
   C. Annotation Families
   D. Loadable Families

8. The Repeating Detail tool places an array of _____

   A. Annotations
   B. Filled Regions
   C. Detail Views
   D. Detail Components

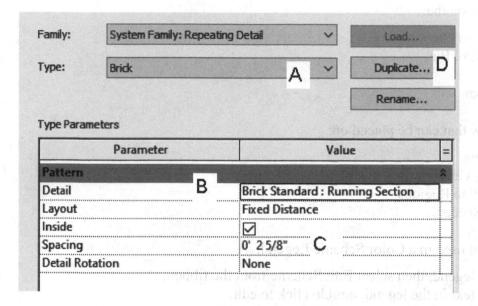

9. Specify the location to select to use a different detail component for the repeating detail.

10. Tagging elements by category:
    A. Allows you to select which elements will be tagged by selecting the element
    B. Allows you to select all the elements in a category (doors/windows/rooms/etc) by simply selecting which tags to be applied in a view
    C. Allows you to select which tag to be placed and then select the element to be tagged one at a time
    D. Applies the category tag to all the selected elements in the entire project

11. T/F Phases allow you to display different versions of a model.

12. T/F Phase filters changes the graphic display of a view depending on the phase.

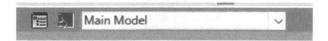

13. This section below the View Control Bar is used to manage:

    A. Phases
    B. Design Options
    C. Worksets
    D. Links

14. T/F Phases can be applied to schedules.

15. To set the Phase to be applied to a view:

    A. Activate the view and select the desired filter in the Properties panel
    B. Go to the Manage tab on the ribbon and set the Phase assigned to each view
    C. Go to the View Template and assign the phase for that view
    D. Go to the View Control Bar and assign the desired phase

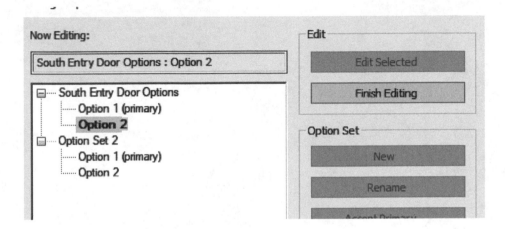

16. You click on Finish Editing. This:

    A. Deletes that design option
    B. Moves elements from one design option to another
    C. Closes the design option you are working in
    D. Incorporates the design option into the main model

ANSWERS: 1) A,B, & D; 2) C; 3) D; 4) C; 5) A.B.C.&D; 6) A,B, & C; 7) D; 8) D; 9) B; 10) A; 11) F; 12) T; 13) B; 14) T; 15) A; 16) C

Notes:

Collaboration

This lesson addresses the following Professional certification exam questions:

- Demonstrate how to copy and monitor elements in a linked file
- Apply interference checking in Revit
- Using Shared Coordinates
- Import DWG, PDF, and Image files
- Assess review warnings
- Worksets
- Linked files

There are no questions regarding collaboration on the user certification exam.

In most building projects, you need to collaborate with outside contractors and with other team members. A mechanical engineer uses an architect's building model to lay out the HVAC (heating and air conditioning) system. Proper coordination and monitoring ensure that the mechanical layout is synchronized with the changes that the architect makes as the building develops. Effective change monitoring reduces errors and keeps a project on schedule.

Worksets are used in a team environment when you have many people working on the same project file. The project file is located on a server (a central file). Each team member downloads a copy of the project to their local machine. The person is assigned a workset consisting of building elements that they can change. If you need to change an element that belongs to another team member, you issue an Editing Request which can be granted or denied. Workers check in and check out the project, updating both the local and server versions of the file upon each check in/out.

Project sharing is the process of linking projects across disciplines. You can share a Revit Structure model with an MEP engineer.

You can link different file formats in a Revit project, including other Revit files (Revit Architecture, Revit Structure, Revit MEP), CAD formats (DWG, DXF, DGN, SAT, SKP), and DWF markup files. Linked files act similarly as external references (XREFs) in AutoCAD. You can also use file linking if you have a project which involves multiple buildings.

It is recommended to use linked Revit models for

- Separate buildings on a site or campus.

- Parts of buildings which are being designed by different design teams or designed for different drawing sets.

- Coordination across different disciplines (for example, an architectural model and a structural model).

Linked models may also be appropriate for the following situations:

- Townhouse design when there is little geometric interactivity between the townhouses.

- Repeating floors of buildings at early stages in the design, where improved Revit model performance (for example, quick change propagation) is more important than full geometric interactivity or complete detailing.

You can select a linked project and bind it within the host project. Binding converts the linked file to a group in the host project. You can also convert a model group into a link which saves the group as an external file.

Exercise 6-1

Monitoring a Linked File

Drawing Name: **i_multiple_disciplines.rvt**
Estimated Time to Completion: 40 Minutes

Scope

Link a Revit Structure file.
Monitor the levels in the linked file.
Reload the modified Structure file.
Perform a coordination review.
Create a Coordination Review report.

Solution

1. Select **Options** on the File menu.

2. In the Username field, type in your first initial and last name.
   Click **OK**.
   Revit will auto-fill the username in reports and the title block.

   Username

   E MOSS

3. If you have an Autodesk online Account, the sign in for the user account is used for your user name. If you don't want to use the name displayed by the Autodesk account, you need to log out of the account.

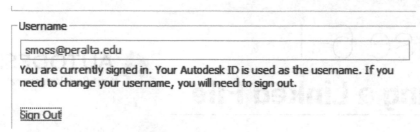

4. Open the *i_multiple_disciplines.rvt* file.

5. 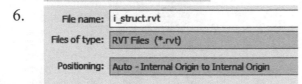 Activate the **Insert** tab on the ribbon.
    Select the **Link Revit** tool on the Link panel.

6. Locate the *i_struct* file.
    Set the Positioning to **Auto-Internal Origin to Internal Origin**.
    Click **Open**.

    File name: i_struct.rvt
    Files of type: RVT Files (*.rvt)
    Positioning: Auto - Internal Origin to Internal Origin

7. Zoom into the left side of the screen where the level markers are.
    There are two sets of level lines. One is part of the original file and the other is from the linked file.

    Roof
    24' - 0"

    Roof TOS
    23' - 4 1/4"

8. If you mouse over the linked file it will show information.

    Roof
    24' - 0"

    Roof TOS
    23' - 4 1/4"

    RVT Links : Linked Revit Model : i_struct.rvt : 4 : location
    <Not Shared>

9. Activate the Collaborate tab on the ribbon.
    Select **Copy/Monitor→Select Link** from the Coordinate panel.
    Pick in the window to select the linked file.

    Copy/ Monitor   Coordination Review   Coordinati Settings
    Use Current Project
    Select Link

10. Select the **Monitor** tool from the Tools panel.

    Options Copy Monitor
    Tools

11. Select the Level 1 level line twice. The first selection will be the host file. The second selection will be the linked file. This is because the level lines overlap.

    Level 1
    0' - 0"

12. You should see a symbol on Level 1 indicating that Level 1 is currently being monitored for changes.

13. 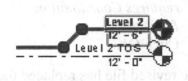 Select the Level 2 level line. The first selection will be the host file. The second selection will be the linked file. You should see a tool tip indicating a linked file is being selected.

14. 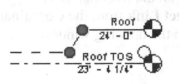 Zoom out. You should see a symbol on Level 2 indicating that Level 2 is currently being monitored for changes.

15. Select the Roof level line.
The first selection will be the host file.
The second selection will be the linked file.
You should see a tool tip indicating a linked file is being selected.

16. Zoom out. You should see a symbol on the Roof level indicating that it is currently being monitored for changes.

17. Warning
Elements already monitored

If you try to select elements which have already been set to be monitored, you will see a warning dialog.
Simply close the dialog and move on.

18. ✓ ✗
Finish Cancel
Copy/Monitor

Select **Finish** on the Copy/Monitor panel.

19.
Manage
Links

Activate the **Insert** tab on the ribbon.
Select **Manage Links** from the Link panel.

20. 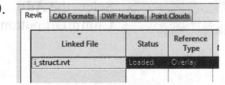 Select the **Revit** tab.
Highlight the *i_struct.rvt* file.

Revit	CAD Formats	DWF Markups	Point Clouds
Linked File	Status	Reference Type	
i_struct.rvt	Loaded	Overlay	

21. Reload From...   Select **Reload From**.

22.  Locate the *i_struct_revised* file.
Click **Open**.

23.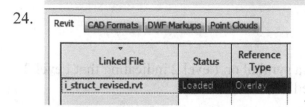
A dialog will appear indicating that the revised file requires Coordination Review.
Click **OK**.

24.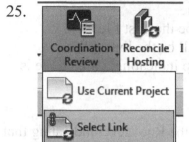
Note that the revised file has replaced the previous link.
This is similar to when a sub-contractor or other consultant emails you an updated file for use in a project.
Click **OK**.

25.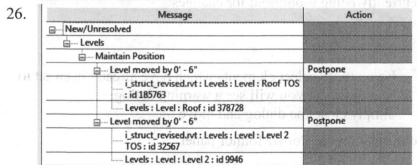
Activate the **Collaborate** tab on the ribbon. Select **Coordination Review→Select Link** from the Coordinate panel. Select the linked file in the drawing window.

26.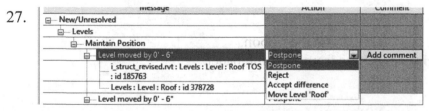
A dialog appears.
Expand the notations so you can see what was changed.

27.
Highlight first change.
Select **Move Level Roof** from the drop-down list.

28.
Select the **Add Comment** button.

29. Type **Approved**.
Click **OK**.

Edit Comment

Approved

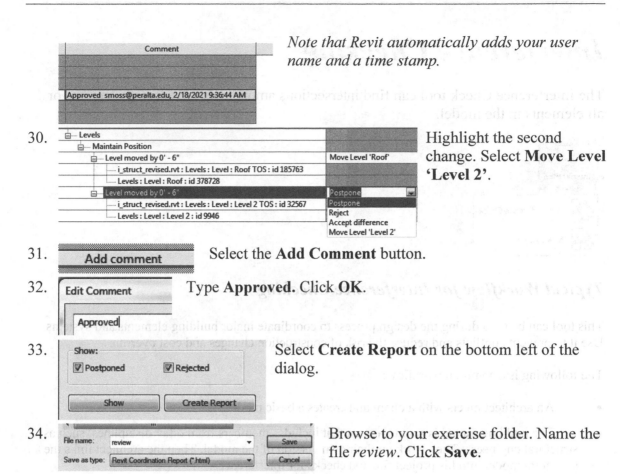

Note that Revit automatically adds your user name and a time stamp.

30.   Highlight the second change. Select **Move Level 'Level 2'**.

31.   Select the **Add Comment** button.

32.   Type **Approved.** Click **OK**.

33.   Select **Create Report** on the bottom left of the dialog.

34. 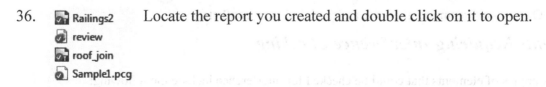   Browse to your exercise folder. Name the file *review*. Click **Save**.

35. Click **OK** to close the Coordination Review dialog box.

If you close the Coordination Review dialog before you create the report and then re-open it to create the report, the report will be blank because all the issues will have been resolved.

36.   Railings2   Locate the report you created and double click on it to open.
      review
      roof_join
      Sample1.pcg

37. This report can be emailed or used as part of the submittal process.

Revit Coordination Report

In host project

New/Unresolved	Levels	Maintain Position	Level moved by 0' - 6"	Levels : Level : Level 2 : id 9946 i_struct_revised.rvt : Levels : Level : Level 2 TOS : id 32567	Approved smoss@peralta.edu, 11/5/2018 1:08:35 PM
New/Unresolved	Levels	Maintain Position	Level moved by 0' - 6"	Levels : Level : Roof : id 378728 i_struct_revised.rvt : Levels : Level : Roof TOS : id 185763	Approved smoss@peralta.edu, 11/5/2018 1:08:17 PM

38. Close the file without saving.

Interference Checking

The Interference Check tool can find intersections among a set of selected elements or all elements in the model.

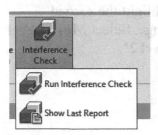

Typical Workflow for Interference Checking

This tool can be used during the design process to coordinate major building elements and systems. Use it to prevent conflicts and reduce the risk of construction changes and cost overruns.

The following is a common workflow:

- An architect meets with a client and creates a basic model.

- The building model is sent to a team that includes members from other disciplines, such as structural engineers. They work on their own version of the model. Then the architect links the structural model into his project file and checks for interferences.

- Team members from other disciplines return the model to the architect.

- The architect runs the Interference Check tool on the existing model.

- A report is generated from the interference check, and undesired intersections are noted.

- The design team discusses the interferences and creates a strategy to address them.

- One or more team members are assigned to fix any conflicts.

Elements Requiring Interference Checking

Some examples of elements that could be checked for interference include the following:

- Structural girders and purlins

- Structural columns and architectural columns

- Structural braces and walls

- Structural braces, doors, and windows

- Roofs and floors

- Specialty equipment and floors

- A linked Revit model and elements in the current model

Exercise 6-2

Interference Checking

Drawing Name: **c_interference_checking.rvt**
Estimated Time to Completion: 15 Minutes

Scope
Describe Interference Checks.
Check and fix interference conditions in a building model.
Generate an interference report.

Solution

1.    If this dialog comes up when you open the file, click **OK**.

2. 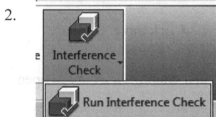   Activate the **Collaborate** tab on the ribbon.
Select the **Interference Check→ Run Interference Check** tool on the Coordinate panel.

3. 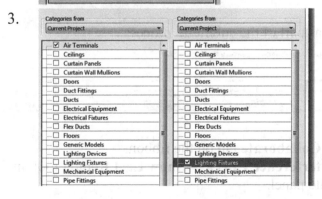   Enable **Air Terminals** in the left panel.
Enable **Lighting Fixtures** in the right panel.

If you select all in both panels, the check can take a substantial amount of time depending on the project and the results you get may not be very meaningful.

4. Click **OK**.

5. Highlight the Lighting Fixture in the first error.

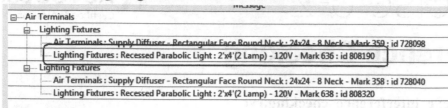

6. 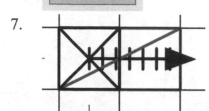 Click the **Show** button.

7. 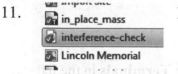 The display will update to show the interference between the two elements.

8. Select **Export**.

9. Browse to your exercise folder.
   Name the file **interference check.**

   Click **Save**.

10. Click **Close** to close the dialog box.

11. 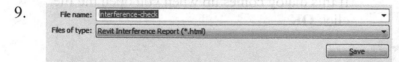 Locate the report you created and double click on it to open.

12. This report can be emailed or used as part of the submittal process.

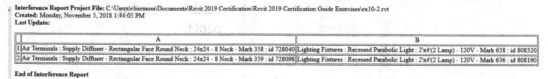

13. Activate the **Collaborate** tab on the ribbon.
    Select the **Interference Check→ Show Last Report** tool on the Coordinate panel.

14. Highlight the first lighting fixture. Verify that you see which fixture is indicated in the graphics window. Close the dialog.

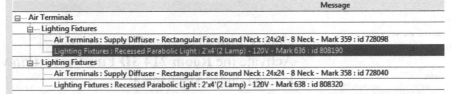

Message
⊟── Air Terminals
⊟── Lighting Fixtures
──── Air Terminals : Supply Diffuser - Rectangular Face Round Neck : 24x24 - 8 Neck - Mark 359 : id 728098
──── Lighting Fixtures : Recessed Parabolic Light : 2'x4' (2 Lamp) - 120V - Mark 636 : id 808190
⊟── Lighting Fixtures
──── Air Terminals : Supply Diffuser - Rectangular Face Round Neck : 24x24 - 8 Neck - Mark 358 : id 728040
──── Lighting Fixtures : Recessed Parabolic Light : 2'x4' (2 Lamp) - 120V - Mark 638 : id 808320

15. 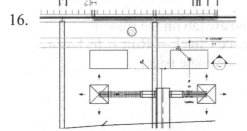 Move the lighting fixture to a position above the air terminal.

Arrange the lighting fixtures and the air diffuser so there should be no interference.

16. 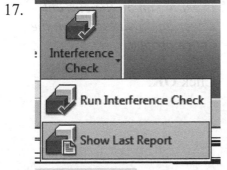 Move the second lighting fixture on the right so it is no longer on top of the air terminal.

Rearrange the fixtures in the room to eliminate any interference.

17. Activate the **Collaborate** tab on the ribbon.
Select the **Interference Check→ Show Last Report** tool on the Coordinate panel.

18. Select **Refresh**.

19. The message list is now empty. Close the dialog box.

20. Highlight **Views** in the Project Browser.
Right click and select **Search**.

21.

Type **Room 214**.
Click **Next**.

22.

Click **Close** when the view is located.
Activate the **Room 214 3D Fire Protection view** located under Mechanical.

23.

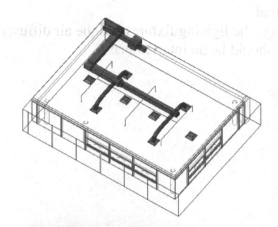

Zoom in so you can see the room.
Window around the entire room so everything is selected.

24.

Activate the **Collaborate** tab on the ribbon.
Select the **Interference Check→ Run Interference Check** tool on the Coordinate panel.

25.

Note that the interference check is being run on the current selection only and not on the entire project.

Click **OK**.

26.

Group by:	Category 1, Category : ▾

Message

- Ducts
  - Pipes
    - Ducts : Round Duct : Taps - Mark 1746 : id 613000
    - Pipes : Pipe Types : Standard - Mark 953 : id 736960
  - Pipes
    - Ducts : Round Duct : Taps - Mark 1752 : id 613012
    - Pipes : Pipe Types : Standard - Mark 952 : id 736959
  - Pipes
    - Ducts : Rectangular Duct : Mitered Elbows / Tees - Mark 1754 : id 613170
    - Pipes : Pipe Types : Standard - Mark 970 : id 736995
  - Pipes
    - Ducts : Rectangular Duct : Mitered Elbows / Tees - Mark 1740 : id 612990
    - Pipes : Pipe Types : Standard - Mark 967 : id 736988

Review the report.
Click **Close**.

27. Close all files without saving.

In the certification exam, you might be asked to identify the id number of the element affected by interference. Can you locate the id number for the element with Mark Number 970 in the report?

Shared Coordinates

Coordinates in a file will only be shared, or the files will have shared coordinates after a process of transferring the coordinate system used in a file into another. It does not matter if in two files the location is exactly the same. They will be not sharing coordinates unless the sharing coordinates process has been carried out.
Normally there is one file, and only one, that is the source for sharing coordinates. From this file the location of different models is transferred to them, and after that all files will be sharing coordinates, and be able to be linked with the "Shared Coordinates option".

Revit uses three coordinate systems. When a new project is opened, the three systems overlap.
They are:
- The Survey Point
- The Project Base Point
- The Internal Origin

The Survey Point is represented by a blue triangle with a small plus symbol in the center. This is the point that stores the universal coordinate system, or a defined global system of the project to which all the project structures will be referred. The Survey Point is a real-world relation to the Revit model. It represents a specific point on the

Earth, such as a geodetic survey marker or a point of reference based on project property lines. The survey point is used to correctly orient the project with other coordinate systems, such as civil engineering software.

- It is the origin that Revit will use in case of share coordinates between models.
- The origin point used when inserting linked files with the Shared Coordinates option.
- The origin point used when exporting with Shared Coordinates option.
- The origin point to which spot coordinates and spot elevations are referenced, if the Survey Point is the coordinate origin in the type properties.
- The origin point to which level elevation is referred, if the SP is the Elevation Base in the level type properties.
- When the view shows the True North, it is oriented according to the Survey Point settings.

The Project Base Point is represented by a blue circle with an x. The position of this point is unique for each model, and this information is not shared between different models. The Project Base point could be placed in the same location as the Survey Point, but it is not usual to work in that way. This point is used to create a reference for positioning elements in relation to the model itself. By default, the Project Base Point is the origin (0,0,0) of the project. The Project Base Point should be used as a reference point for measurements and references across the site. The location of this point does not affect the shared site coordinates, so it should be located in the model where it makes sense to the model, usually at the intersection of two gridlines or at the corner of a building.

- The origin point to which spot coordinates and spot elevations are referenced, if the Project Base Point is the coordinate origin in the type properties.
- The origin point to which level elevation is referred, if the Project Base Point is the Elevation Base in the level type properties.
- When the view shows the Project North, it is oriented according to the Project Base Point.
- If it is not necessary, it is better not to move this point, so that it always is coincident with the third point: the Internal Origin.
- If we have moved the Project Base Point to a different location, it can be placed back to the initial position by right clicking on it when selected, and use **Move to Startup Location.**

The Internal Origin is now visible starting with the 2022 release of Revit. The Internal Origin is coincident with the Survey Point and the Project Base Point when a new project is started. It is displayed with an XY icon. Prior to this release, users sometimes resorted to placing symbols at the internal origin location to keep track of the origin.

- The project should be modeled in a restricted area around the internal origin point. The model should be inside a 20 miles radius circle around the internal origin, so that Revit can compute accurately.

- This is the origin point that is used when inserting external files using the "Internal Origin to Internal Origin" option.

- This is the origin point that is used when copy/pasting model objects from one file into another using the "aligned" option.

- The origin point to which spot coordinates and spot elevations are referenced, if Relative is the coordinate origin in the type properties.

- The origin point used when exporting with Internal Coordinates option.

- Revit API and Dynamo use this point as the coordinates origin point for internal computational calculations.

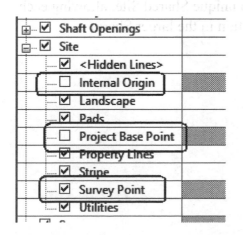

The visibility of these three systems is controlled under the Site category on the Model tab of the Visibility Graphics dialog.

Moving the project base point is like changing the global coordinates of the entire project.

Another way of looking at it is—if you modify the location of the project base point, you are moving the model around the earth; if you modify the location of the survey point, you are moving the earth around the model.

The benefits of correctly adjusting shared coordinates are not limited to aligning models for coordination but can also position the project in the real world if using proper geodetic data from a surveyor landmark or a provided file with the real-world information present.

To fully describe where an object (a building) sits in 3D space (its location on the planet), we need four dimensions:

- East/West position (the X coordinate)
- North/South position (the Y coordinate)
- Elevation (the Z coordinate)
- Rotation angle (East/West/True North)

These four dimensions uniquely position the building on the site and orients the building relative to a known benchmark as well as other landmarks. Revit allows you to assign a unique name to these four coordinates. Revit designates this collection as a **Shared Site**. You can designate as many Shared Sites as you like for a project. This is useful if you are planning a collection of buildings on a campus. By default, every project starts out with a single Shared Site which is named Internal. The name indicates that the Shared Site is using the Internal Coordinate System. Most projects will only require a single Shared Site, but if you are managing a project with more than one building or construction on the same site, it is useful to define a Shared Site for each instance of a linked file.

For example, in a resort, there may be different instances of the same cabin located across the site. You can use the same Revit file with the cabin model and copy the link multiple times. Each cabin location will be assigned a unique Shared Site, allowing each instance of the linked file to have its own unique location in the larger Shared Coordinate system.

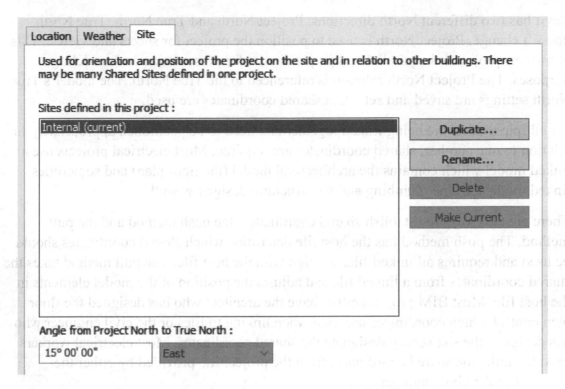

Shared Sites are named using the Location Weather and Site dialog.

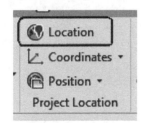

The dialog is opened using the Location tool on the Manage tab on the ribbon.

Why setting up shared coordinates is important:

- To properly coordinate the project models to work across platforms, including BIM Track.
- To display the project in its real-world location (including BIM and GIS data overlay).
- To link files together that have different base points (arbitrary coordinate systems).

Shared Coordinates allow you to adjust the reference point from project origin to a shared site point, to make it appear in a different position or with a different angle based on the internal base point. It is important to note that nothing **in the Revit model moves when using this system,** even if it looks like it moves and/or rotates on the screen. It is just applying an adjustment to the point of origin, so in fact, all elements keep their relation to the internal base point when using this option.

Revit has two different North directions: Project North and True North. True North doesn't change. Project North is used to position the project for sheets and views. This allows the project to be parallel and perpendicular for presentation and modeling purposes. The Project North rotation is referenced to the True North. The model's True North settings are saved and set when shared coordinates are used.

If multiple models are being linked together and need to be positioned appropriately in relation to one another, shared coordinates are required. Most electrical projects use a linked model which contains the architectural model (the floor plan) and sometimes linked models for the plumbing and the structural designs as well.

There are two ways to establish shared coordinates: the push method and the pull method. The push method has the host file determine which shared coordinates should be used and requires all linked files to align with the host file. The pull method takes the shared coordinates from a linked file and adjusts the position of the model elements in the host file. Most BIM projects either have the architect who has designed the floor plan control which coordinates are used when linking to files or the civil engineer who has designed the site survey designate the shared coordinates. Most electrical workers need to "pull" the shared coordinates from the project file provided by either the architect or the civil engineer.

The steps to pull shared coordinates are as follows:

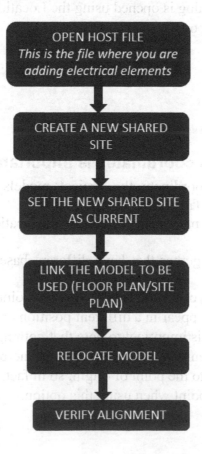

Exercise 6-3

Understanding Shared Coordinates

Drawing Name: new
Estimated Time: 5 minutes

Scope:
- ❑ Shared Coordinate System
- ❑ Site Point
- ❑ Base Point
- ❑ Internal Origin
- ❑ Views
- ❑ Visibility/Graphics

Solution

1.

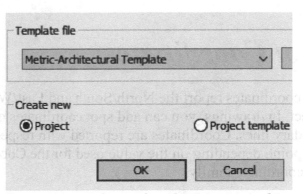

Start a new project using the *Metric-Architectural* template.

2.

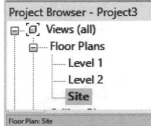

Open the Site floor plan.

3.

Set the Discipline to **Coordination**.

4.

Type **VV** to open the Visibility/Graphics dialog.

On the Model Categories tab:

Enable:
- Internal Origin
- Project Base Point
- Survey Point

These are located under the Site category.

Click **OK**.

The Internal Origin, Project Base Point and Survey Point are all visible.

They are overlaid on top of each other.

5.

Save as *ex6-3.rvt*.

Spot Coordinates

Spot coordinates report the North/South and East/West coordinates of points in a project. In drawings, you can add spot coordinates on floors, walls, toposurfaces, and boundary lines. Coordinates are reported with respect to the survey point or the project base point, depending on the value used for the Coordinate Origin type parameter of the spot coordinate family.

Exercise 6-4

Placing a Spot Coordinate

Drawing Name: spot_coordinate.rvt.
Estimated Time: 20 minutes

Scope:
- ❏ Spot Coordinate
- ❏ Specify Coordinates at a Point
- ❏ Report Shared Coordinates

Solution:

We can add a spot coordinate symbol to the project base point to keep track of its location.

1. 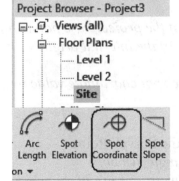   Open the Site floor plan.

2. 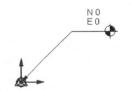   Switch to the Annotate tab on the tab on the ribbon.

    Select the **Spot Coordinate** tool on the Dimension panel.

3.   Select the Project Base Point.

    Left click to place the spot coordinate.
    That is the circular symbol.

    Cancel out of the command.

4.

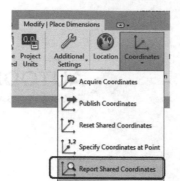

Switch to the Manage tab on the tab on the ribbon.

Select **Report Shared Coordinates** on the Project Location panel.

Select the Project Base Point.

5.

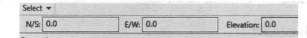

The coordinates are displayed on the Options bar.

Cancel out of the command.

6.

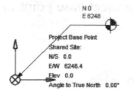

Select the Project Base Point.

Notice that the coordinates that are displayed are blue. This means they are temporary dimensions which can be edited.

7.

Click on the **E/W** dimension.
Change it to **6248.4.**
Click **ENTER.**

8.

The location of the project base point shifted to be 6248.4 m east of the internal origin/survey point.

Notice that the spot coordinate value updated.

9.

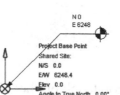

Left click on the Survey Point.
That's the triangle symbol.
Notice that dimensions are black, which means they cannot be edited.

10.

Drag the survey point away from the internal origin (the XY icon).

Notice that the Project Base Point value updates.

Remember the Survey Point represents a real and known benchmark location in the project (usually provided by the project survey). The Project Base Point is simply a known point on the building (usually chosen by the project team).

Think of the Survey Point as the coordinates in the *World* and the Project Base Point as the local building coordinates.

The Survey Point is always located at 0,0, while the Project Base Point is relative to the Survey Point.

It is called a shared coordinate system because it represents the coordinate system of the world around us and is shared by all the buildings on the site.

11.

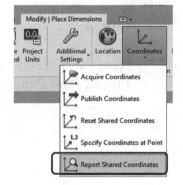

Select **Report Shared Coordinates** on the Project Location panel.

Select the Internal Origin (the XY icon).

12.

The coordinates are displayed on the Options bar.

The values will be different from mine.

Notice that the Internal Origin is located relative to the Survey Point. The Internal Coordinate System represents the building's "local" coordinates.

Cancel out of the command.

13. 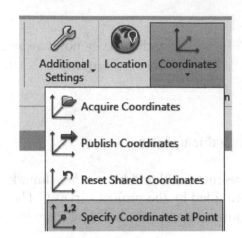 *Usually, you will be provided the coordinate information for a project from the civil engineer.*

Select **Specify Coordinates at Point**.

Select the Survey Point (the triangle symbol.).

14.

Relocate this project in Shared Coordinates by specifying known values at the point you selected. Current project will move relative to globally positioned links.

New Coordinates

North/South:	11988.8
East/West:	4876.8
Elevation:	6096.0

Angle from Project North to True North

15° 00' 00"	East

| OK | Cancel |

Fill in the Shared Coordinates:

For North/South: **11988.8.**
For East/West: **4876.8.**
For Elevation: **6096.**
For Angle from Project North to True North: **15° East.**

Click **OK**.

15.

Zoom out and you will see that the Survey Point's position has shifted.

Note the coordinates for the Project Base Point updated.

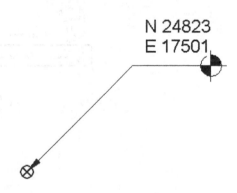

N 24823
E 17501

16. Save as *ex6-4.rvt*.

Exercise 6-5

Understanding Location

Drawing Name: Simple Building.rvt
Estimated Time: 20 minutes

Scope:

- ❑ Linking Files
- ❑ Shared Coordinate System
- ❑ Survey Point
- ❑ Base Point
- ❑ Internal Origin
- ❑ Visibility/Graphics
- ❑ Spot Elevation

Solution:

1.

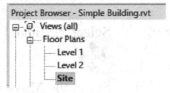

Open the Site floor plan.

2.

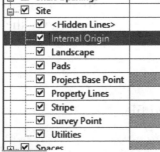

The Survey Point and the Project Base Point are visible, but not the Internal Origin.

Open the Visibility/Graphics dialog by typing **VV**.

3.

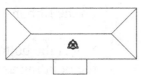

On the Model Categories tab:

Enable: **Internal Origin**

This is located under the Site category.

Click **OK** and close the dialog.

4.

Activate the **Insert** tab on the ribbon.

Select **Link Revit**.

5.

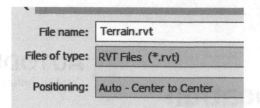

Locate the *Terrain.rvt* file.

Set the Positioning to **Auto – Center to Center**.

Click **Open**.

Many users will be tempted to align the files using Internal Origin to Internal Origin. This is not the best option when we plan to reposition the building on the site plan. Origin to Origin works well when linking MEP files to the Architectural file.

Notice that the linked file's coordinate system is slightly offset from the host file's coordinate system.

Next, we will re-position the linked file into the correct relative position. If we do it this way, we are not moving the building, we are moving the terrain. We will still use the terrain to establish the Shared Coordinate system, which is the "pull" method. The "pull" method pulls the information from the linked file.

6.

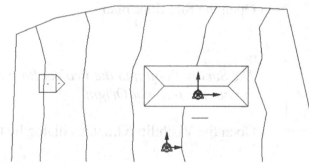

Select the Move tool on the tab on the ribbon.
Select the bottom midpoint of the building as the start point.
Move the cursor to the left and type 15' to move the terrain left 15'.

Select the linked file.

Select the **Move** tool on the tab on the ribbon.

Select the bottom midpoint of the building as the start point.

Move the cursor down and type 30' to move the terrain down 30'.

7.

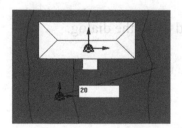

Select the linked file.

Select the Rotate tool on the tab on the ribbon.

Start the angle at the horizontal 0° level.
Rotate up and type **20°**.

8.

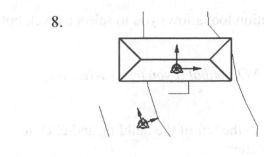

We have repositioned the terrain on the XY plane, but we have not changed the elevation.

Click on the Project Base Point on the building.

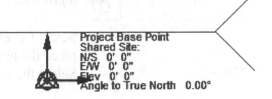

Notice that it has not changed.

9.

Click on the Project Base Point for the terrain file.

It shows as 0,0 still as well.

10.

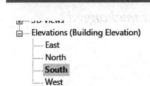

Open the **South** elevation.

11.

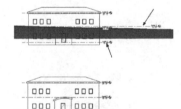

Notice that the levels are not aligned between the files.

12.

*Use the **ALIGN** tool to move the Level 1 in the terrain file to align with Level 1 in the simple building file.*

Select the **ALIGN** tool.

Select the Level 1 that is on the simple building file as the target.
Select the Level 1 on the terrain file to be moved.

Cancel out of the ALIGN command.

13.

Spot
Elevation

Switch to the Annotate tab on the ribbon.

Select **Spot Elevation**.

14.

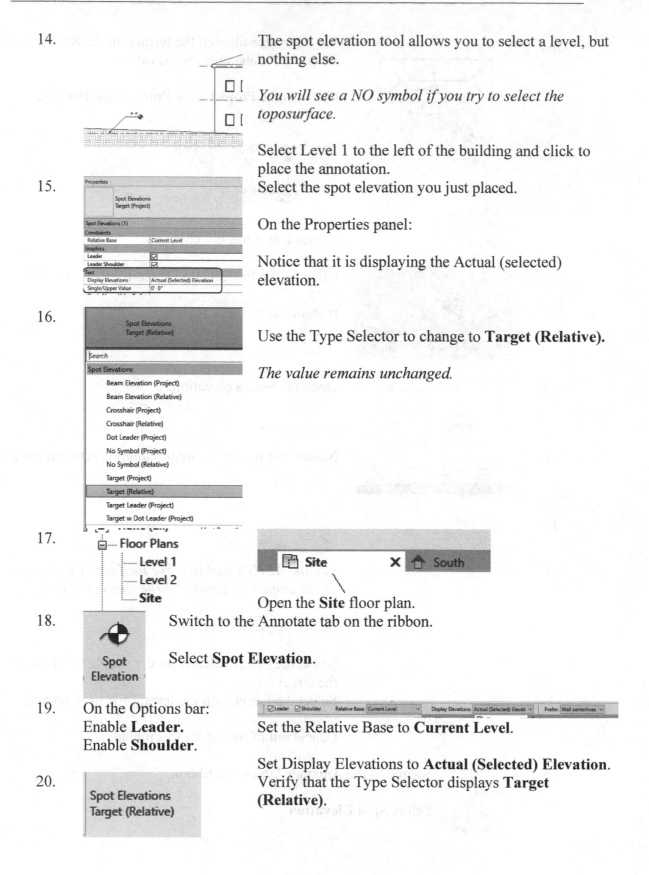

The spot elevation tool allows you to select a level, but nothing else.

You will see a NO symbol if you try to select the toposurface.

Select Level 1 to the left of the building and click to place the annotation.
Select the spot elevation you just placed.

15.

On the Properties panel:

Notice that it is displaying the Actual (selected) elevation.

16.

Use the Type Selector to change to **Target (Relative).**

The value remains unchanged.

17.

Open the **Site** floor plan.

18.

Switch to the Annotate tab on the ribbon.

Select **Spot Elevation**.

19. On the Options bar:
Enable **Leader.**
Enable **Shoulder**.

Set the Relative Base to **Current Level**.

20.

Set Display Elevations to **Actual (Selected) Elevation**.
Verify that the Type Selector displays **Target (Relative)**.

21. 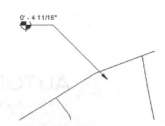 Click on the toposurface.

The elevation is displayed in black.
This means it cannot be modified. The value is
generated by the actual location of the toposurface.

The value you see may be different.

22.  Verify that both site points display 0,0 as their
coordinates.

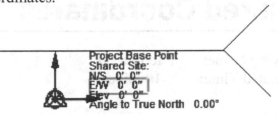

23. 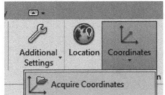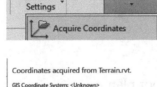 Switch to the Manage tab on the tab on the ribbon.

Select **Acquire Coordinates** and then select the linked
terrain file.

24. Coordinates acquired from Terrain.rvt.

GIS Coordinate System: <Unknown>

Close

Click **Close**.

25. 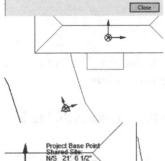 The Survey Point in the host file (simple building)
moved. It is now aligned with the Survey Point on the
linked file (terrain).

26. 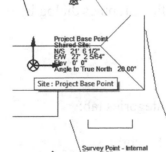 Select the Project Base Point for the simple building.

Notice the change to the coordinates.

27. 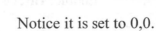 Select the Survey Point.

Notice it is set to 0,0.
The spot elevation value is unchanged because the
coordinates were acquired from the linked file and the
spot elevation is reporting the elevation of the linked
file.

28. Save the file as *ex6-5.rvt*.

Exercise 6-6

Linking Files using Shared Coordinates

Drawing Name: terrain.rvt
Estimated Time: 10 minutes

Scope:
- ❑ Linking Files
- ❑ Shared Coordinate System
- ❑ Survey Point
- ❑ Base Point
- ❑ Internal Origin
- ❑ Visibility/Graphics

Solution

1.
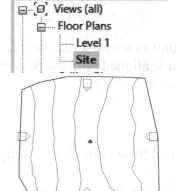

Open the Site floor plan.

2.

The Survey Point and the Project Base Point are visible, but not the Internal Origin.

Open the Visibility/Graphics dialog by typing **VV**.

3.
☐ ☑ Site
☑ <Hidden Lines>
☑ Internal Origin
☑ Landscape
☑ Pads
☑ Project Base Point
☑ Property Lines
☑ Stripe
☑ Survey Point
☑ Utilities
☐ ☑ Spaces

On the Model Categories tab:

Enable: **Internal Origin**

This is located under the Site category.

Click **OK** and close the dialog.

4.

Use **FILTER** to select just the Survey Point.

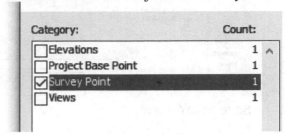

Notice that it is still located at 0,0.

Click ESC to release the selection.

5.

Select the Project Base Point.

It is also located at 0,0.

The file wasn't changed because we acquired the coordinates from the linked file and applied them to the host file.

6.

Switch to the **Insert** tab on the tab on the ribbon.

Select **Link Revit**.

7.

File name: Simple Building 2.rvt

Files of type: RVT Files (*.rvt)

Positioning: Auto - By Shared Coordinates

Select *Simple Building 2.rvt*.
Set the Positioning to **Auto -By Shared Coordinates.**

Simple Building 2 has the coordinates assigned that we defined in the previous exercise.

Click **Open**.

8.

The linked file is positioned in the correct location.

9.

Click on the building.
Notice that Project Base Point is using the Shared Site coordinates.

10.

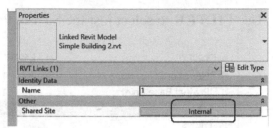

With the linked file selected, notice in the Properties palette, it is using the Internal Shared Site definition.

11.   Save as *ex6-6.rvt*.

Exercise 6-7

Defining a Shared Site

AUTODESK.
Certified Professional

Drawing Name: terrain 2.rvt
Estimated Time: 30 minutes

Scope

- ❑ Linking Files
- ❑ Shared Coordinate System
- ❑ Survey Point
- ❑ Base Point
- ❑ Shared Site

Solution

1.
Project Browser - Building A.rvt
  Views (all)
    Floor Plans
      Level 1
      Level 2
      Site

Open Building A.
Open the Site floor plan.

2.

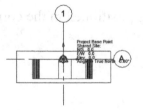

Verify that the Survey Point and Base Point are located at 0,0.

Close the file.

3.
Floor Plans
  Platform 1
  Platform 2
  Platform 3
  Site (Dimensioned)
  Site (Project North)
  Site (True North)
  Site Datum

Open the *terrain 2. rvt* file.

Open the **Site (Project North)** floor plan.

4. The site plan has building pads for three buildings at three different locations on the site.

5. Switch to the **Insert** tab on the tab on the ribbon.

   Select **Link Revit**.

6. Select Building A.*rvt.*
   Set the Positioning to **Auto -Center to Center.**

File name:	Building A.rvt
Files of type:	RVT Files (*.rvt)
Positioning:	Auto - Center to Center

   Click **Open**.

7. *Note the position of Building A on the site plan.*

   Use the **ALIGN** tool to move Building A onto the A platform. Align Grids 1 to each other and Grids A to each other.

   Remember to select the Platform grid first as the target and then the building grid.

8. Open the South (Full) elevation view.

   - {3D}
   - Elevations (Building Elevation)
     - East
     - North
     - South (Cropped)
     - **South (Full)**
     - West

9. *Notice that the Building A elevation needs to be adjusted.*

10. Use the **ALIGN** tool to align Level 1 for Building A to Platform 1.

11. Open the **Platform 1** view.

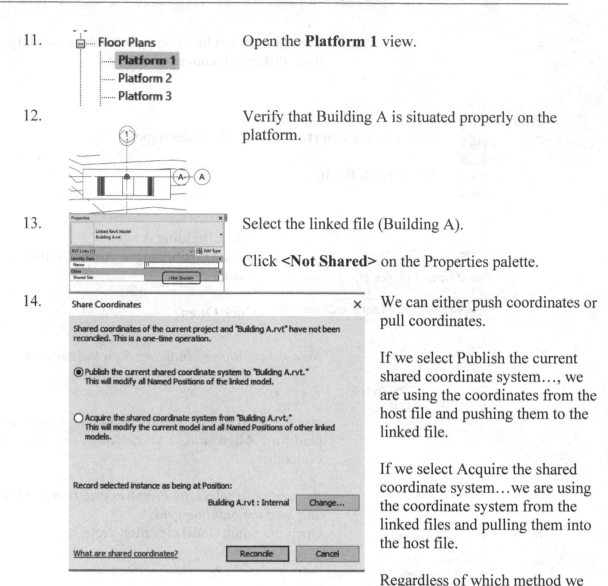

12. Verify that Building A is situated properly on the platform.

13. Select the linked file (Building A).

Click **<Not Shared>** on the Properties palette.

14. We can either push coordinates or pull coordinates.

If we select Publish the current shared coordinate system…, we are using the coordinates from the host file and pushing them to the linked file.

If we select Acquire the shared coordinate system…we are using the coordinate system from the linked files and pulling them into the host file.

Regardless of which method we use, after the selection, they will be sharing the same coordinate system.

Since we started with the terrain file, it is standard to let the topo file control which coordinate system is used in a project.

Remember that whichever option you select does not change the linked file or the host file local internal coordinates. We are only defining how we want the files to interact relative to each other.

15.

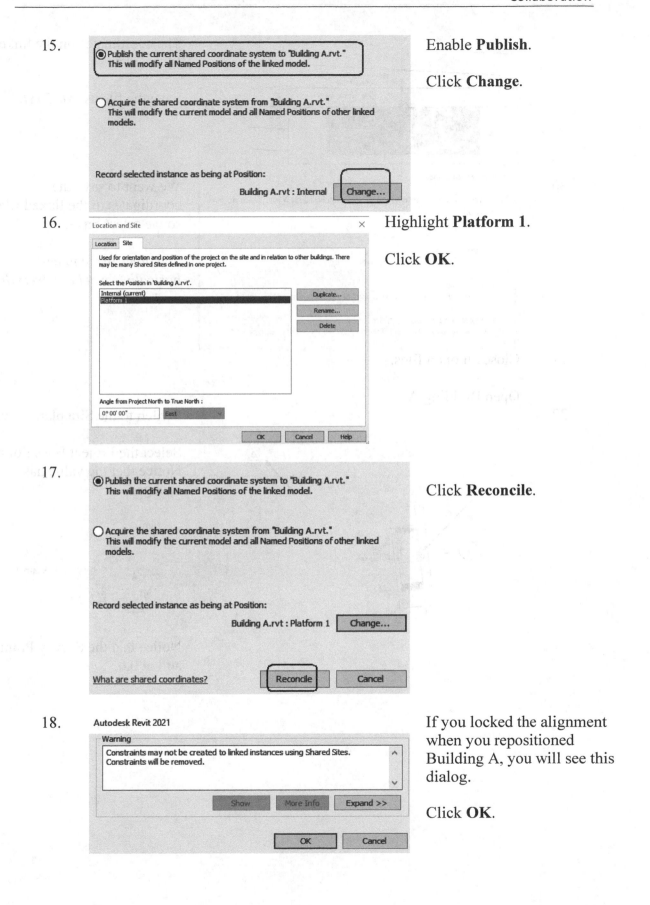

Enable **Publish**.

Click **Change**.

16.

Highlight **Platform 1**.

Click **OK**.

17.

Click **Reconcile**.

18.

If you locked the alignment when you repositioned Building A, you will see this dialog.

Click **OK**.

19.

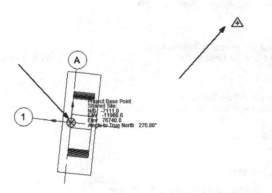

The coordinates on the linked files update.

Save the file as *ex6-7.rvt*.

20.

Location Position Changed

You have changed the "current" Position in Building A.rvt. What do you want to do?

→ Save
Saves the new position back to the link.

→ Do not save
Returns to the previously saved position when the link is reloaded or reopened.

→ Disable shared positioning
Retains the current placement of the link and clears the Shared Position parameter.

We want to save the coordinates to the linked file, so we select **Save**.

The other two choices basically cancel out everything we just did.

21.   Close all open files.

       Open Building A.

22.

Switch to the Site plan view.

Select the Project Base Point. Notice that the value has changed.

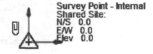

Notice that the Survey Point is still at 0,0.

23.

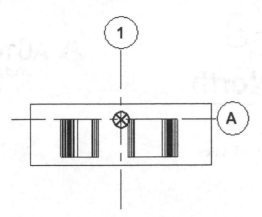

Open the Level 1 view.

The building is oriented horizontally and vertically on the screen to make it easy to work on.

24. Open the **Site** plan.

├─ Floor Plans
│   ├─ Level 1
│   ├─ Level 2
│   └─ **Site**

25.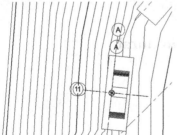

Switch to the **Insert** tab on the tab on the ribbon.

Select **Link Revit**.

26.

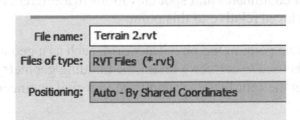

File name:	Terrain 2.rvt
Files of type:	RVT Files (*.rvt)
Positioning:	Auto - By Shared Coordinates

Select *terrain 2.rvt*.
Set the Positioning to **Auto - By Shared Coordinates.**

Click **Open**.

27. Notice that the terrain is oriented perfectly to the building.

This is because we are using the same shared coordinates between the two files.

Close all files.

Exercise 6-8
Using Project North

AUTODESK.
Certified Professional

Drawing Name: **Import Site.Dwg**
Estimated Time to Completion: 40 Minutes

Scope

Link an AutoCAD file.
Set Shared Coordinates.
Set Project North.

Every project has a project base point ⊗ and a survey point △, although they might not be visible in all views because of visibility settings and view clippings. They cannot be deleted.

The project base point defines the origin (0,0,0) of the project coordinate system. It also can be used to position the building on the site and for locating the design elements of a building during construction. Spot coordinates and spot elevations that reference the project coordinate system are displayed relative to this point.

The survey point represents a known point in the physical world, such as a geodetic survey marker. The survey point is used to correctly orient the building geometry in another coordinate system, such as the coordinate system used in a civil engineering application.

Solution

1. Start a new project file using the **Metric-Architectural** template.

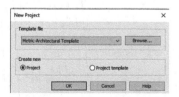

2. Activate the **Site** floor plan.

3. Activate the **Insert** tab on the ribbon.
   Select the **Link CAD** tool on the Link panel.

4. Select the *Import Site* drawing. Set Colors to **Preserve**. Set Import Units to **Auto-Detect**. Set Positioning to **Auto - Center to Center**. Click **Open**.

File name:	Import Site
Files of type:	DWG Files (*.dwg)

Colors: Preserve Positioning: Auto - Center to Center
Layers/Levels: All Place at: Level 1
Import units: Auto-Detect 1.000000 ☑ Orient to View
☑ Correct lines that are slightly off axis Open Cancel

5. Name | Import Site.dwg
Other
Shared Site | <Not Shared>

Select the imported site plan. In the Properties pane, select the Shared Site button.

6. ⦿ Acquire the shared coordinate system from "Import Site.dwg." This will modify the current model and all Named Positions of other linked models.

Record selected instance as being at Position:
Import Site.dwg : DefaultLocation Change...

Enable **Acquire the shared coordinate system…** Select **Change**.

7. Location Weather Site
Define Location by:
Internet Mapping Service
800 Fallon St, Oakland, CA 94607 Search

OK Cancel Help

Activate the Location tab. Enter the Project Address. Click **Search** and then click **OK**.

You must be connected to the internet in order to input an address.

8. Reconcile Click **Reconcile**.

9.

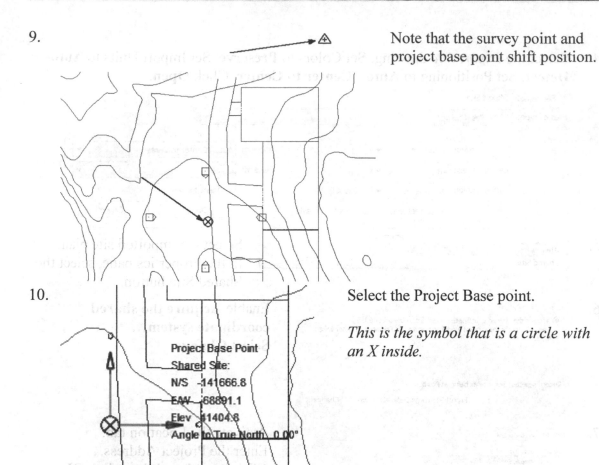

Note that the survey point and project base point shift position.

10.

Select the Project Base point.

This is the symbol that is a circle with an X inside.

Project Base Point
Shared Site:
N/S   -141666.8
E/W   -68891.1
Elev   41404.8
Angle to True North   0.00°

11.

Project Base Point (1)	
Identity Data	
N/S	-141666.8
E/W	-68891.1
Elev	41404.8
Angle to True North	90.00°

Change the Angle to Truth North to **90.00**.
Click **Apply**.
The site plan will rotate 90 degrees.

12.

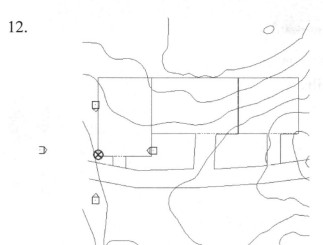

Move the import site plan so that the corner of the rectangle is coincident with the project base point.
Use the **MOVE** tool.
Select the corner of the building and then the Project Base Point.

13.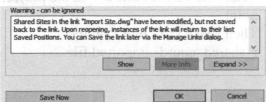

    Select **OK**.

    If you select Save Now, this will modify the linked file.

14.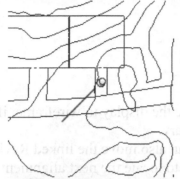

    Use the PIN tool to prevent the site drawing from shifting around.

    Select the linked file first, then select the PIN.

15.

    Activate the Insert tab on the ribbon.
    Select **Link Revit** from the Link panel.

16.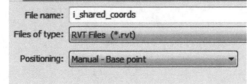

    Select the *i_shared_coords* file.
    Set the Positioning to **Manual - Base point**.
    Click **Open**.

17.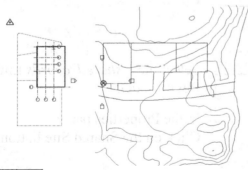

    Place the Revit file to the left of the site plan.

18.   Select the **Align** tool on the Modify panel on the Modify tab on the ribbon.

19.

Select the vertical line above the project base point.
Then select Grid line 1 on the i_shared_coords imported file.

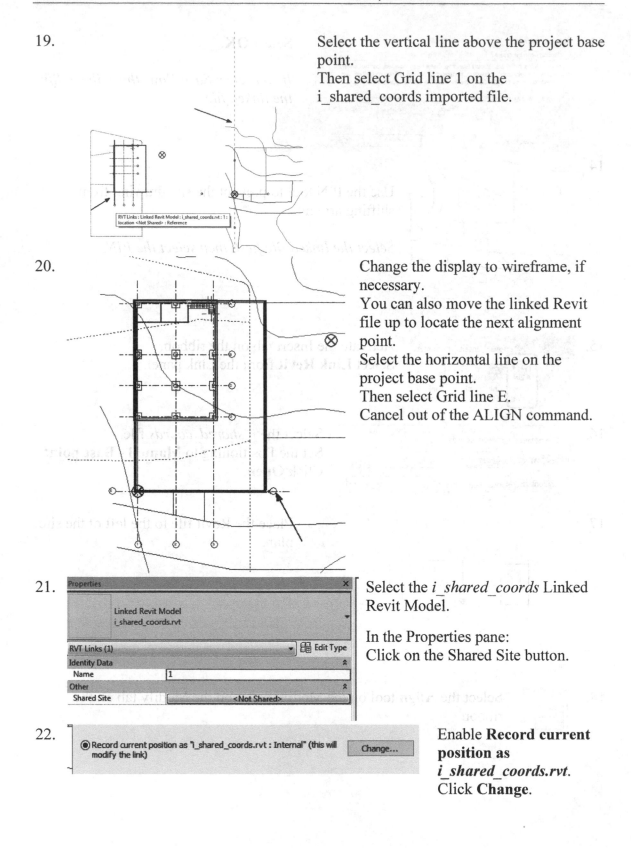

20.

Change the display to wireframe, if necessary.
You can also move the linked Revit file up to locate the next alignment point.
Select the horizontal line on the project base point.
Then select Grid line E.
Cancel out of the ALIGN command.

21.

Select the *i_shared_coords* Linked Revit Model.

In the Properties pane:
Click on the Shared Site button.

22.

Enable **Record current position as i_shared_coords.rvt**.
Click **Change**.

23. 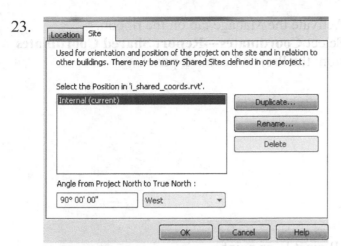 Click **OK**.
Click **OK** to close the Select Site dialog.

24. Elevations (Building Elevation)   Activate the North Elevation.
- East
- **North**
- South
- West

25.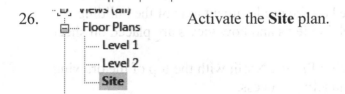

Level 1 and Level 2 are in the host project.
The other levels shown are in the Linked Revit model.

26. Views (all)
- Floor Plans
  - Level 1
  - Level 2
  - **Site**

Activate the **Site** plan.

27. 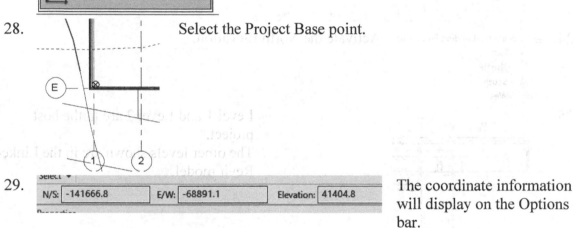 Activate the Manage tab on the ribbon.
Select **Coordinates→Report Shared Coordinates** from the Project Location panel.

28. Select the Project Base point.

29. The coordinate information will display on the Options bar.

| N/S: -141666.8 | E/W: -68891.1 | Elevation: 41404.8 |

30. Close without saving.

Project North

All models have two north orientations: Project North and True North.

- **Project North** is typically based on the predominant axis of the building geometry. It affects how you sketch in views and how views are placed on sheets.

  Tip: When designing the model, align Project North with the top of the drawing area. This strategy simplifies the modeling process.

- **True North** is the real-world north direction based on site conditions.

  Tip: To avoid confusion, define True North only after you begin modeling with Project North aligned to the top of the drawing area and after you receive reliable survey coordinates.

All models start with Project North and True North aligned with the top of the drawing area, as indicated by the survey point △ and the project base point ⊗ in the site plan view.

You may want to rotate True North for the following reasons:

- to represent site conditions
- for solar studies and rendering to ensure that natural light falls on the correct sides of the building model
- for energy analysis
- for heating and cooling loads analysis

Exercise 6-9

△ AUTODESK.
Certified Professional

Using True North vs Project North

Drawing Name: parcel map.rvt
Estimated Time: 20 minutes

Scope

- ❑ Project North
- ❑ True North

Solution

1. 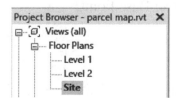 Open the Site floor plan.

2.

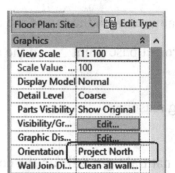

On the Properties panel:

Verify that the Orientation is set to **Project North**.

3.

Select **Model Line** from the Architecture tab on the ribbon.

4.

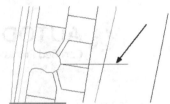

Draw a horizontal line from the east quadrant of the cul de sac to the right.

5.

Place an Angular dimension to determine the orientation of the parcel map.

6.

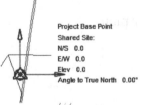

The angle value is 9°.

Delete the model line and the angular dimension.

7.

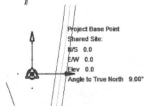

Select the Project Base Point.

8.

Change the Angle to True North to 9°.

9. 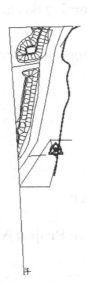 Window around all the elements in the view.

10. Select **Filter** on the ribbon.

Filter

11. Scroll down to the bottom and uncheck:

- Project Base Point
- Survey Point
- Views

Click **OK**.

12. Select **Rotate** on the Modify tab on the ribbon.

13.  Click **Place** next to Center of Rotation on the Options bar.

Disable **Copy**.

14.  Select the Project Base Point.

15. Select a horizontal point to the right of the Project Base Point as the starting point of the angle.

Drag the mouse up to set the end angle at 9°.

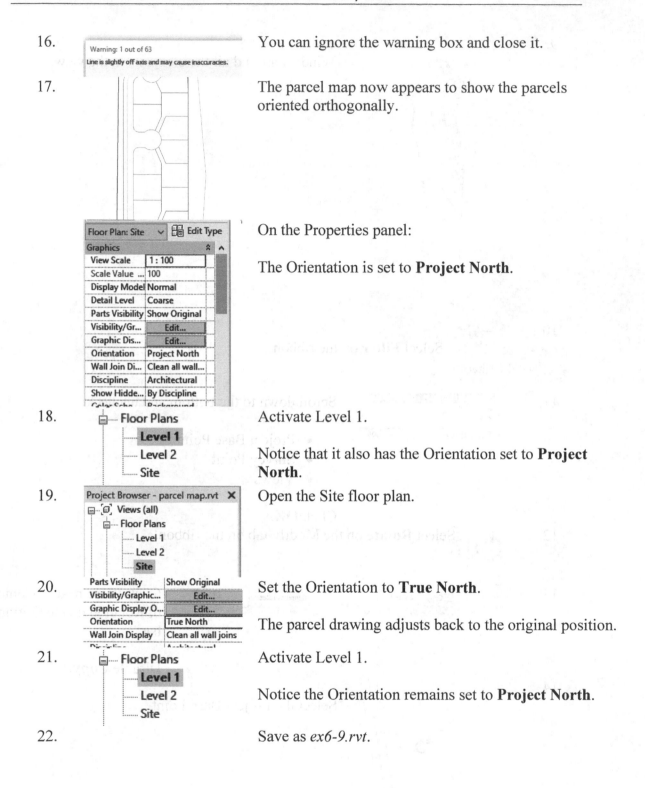

16. You can ignore the warning box and close it.

17. The parcel map now appears to show the parcels oriented orthogonally.

On the Properties panel:

The Orientation is set to **Project North**.

18. Activate Level 1.

Notice that it also has the Orientation set to **Project North**.

19. Open the Site floor plan.

20. Set the Orientation to **True North**.

The parcel drawing adjusts back to the original position.

21. Activate Level 1.

Notice the Orientation remains set to **Project North**.

22. Save as *ex6-9.rvt*.

Importing Files

Revit allows you to add raster images or PDF files into a building model. These can be inserted into 2D views – elevations, floor plans, ceiling plans, and sections.

AutoCAD DWG files can also be imported or linked into a Revit building model.

Linking AutoCAD files allows them to be used as underlays for views. If the AutoCAD DWG is modified, it will automatically update in the Revit file if it is reloaded.

If you import an AutoCAD file, it is inserted into the building model. It will not update if the external file is changed.

Exercise 6-10

Import DWG

Drawing Name: property survey.dwg
Estimated Time: 10 minutes

Scope

- ❑ Import CAD
- ❑ Property Line
- ❑ Toposurface

Solution

1. 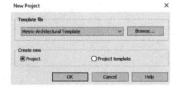   Start a New project using the Metric-Architectural template.

   Click **OK**.

2.    Switch to the Insert tab on the ribbon.

   Select **Import CAD**.

3.

File name: property survey.dwg
Files of type: DWG Files (*.dwg)

Colors: Preserve
Layers/Levels: All
Import units: feet   1.000000
☑ Correct lines that are slightly off axis
Positioning: Auto - Origin to Internal Origin
Place at: Level 1
☑ Orient to View
Open   Cancel

Select the *property survey.dwg*.

Set Colors to **Preserve**.
Set Positioning to **Auto- Origin to Internal Origin.**
Set Layers/Levels to **All.**
Set Import Units to **feet.**
Click **Open.**

4.

Views (all)
Floor Plans
Level 1
Level 2
Site

Open the **Site** plan view.

5.

n Property C
Line

Switch to the Massing & Site tab on the ribbon.

Select **Property Line**.

6.

How would you like to create the property lines?

→ Create by entering distances and bearings

→ Create by sketching

Select **Create by sketching**.

7.

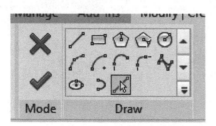

Mode   Draw

Select the Pick Line tool.

Select the elements in the property survey drawing.

8.

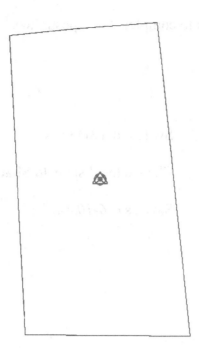

You should see a completed outline.

Select the Green Check on the ribbon to complete.

If you get a warning, zoom in and see if you missed any portions of the outline.

9. Select **Toposurface.**

10.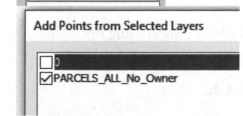

Select **Create from Import →Select Import Instance.**

Select the imported drawing.

11. Enable the **PARCELS_All_No_Own**er layer.

Click **OK**.

12.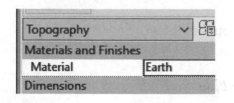

Set the Material to **Earth** in the Properties Panel.

13.  Select **Green Check** to complete the toposurface.

14.

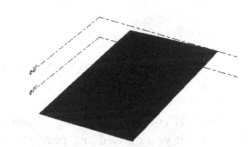

Switch to a 3D view.

Change the display to **Shaded.**

Save as *ex6-10.rvt*.

Exercise 6-11

Import PDF

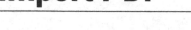

A AUTODESK.
Certified Professional

Drawing Name: bldg._8_2.pdf
Estimated Time: 5 minutes

Scope

❑   Import PDF

Solution

1.  Start a New project using the Metric-Architectural template.

   Click **OK.**

2. Switch to the Insert tab on the ribbon.

   Select **Import PDF.**

3. 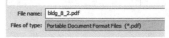 Select *bldg._8_2.pdf.*

   Click **Open.**

4.
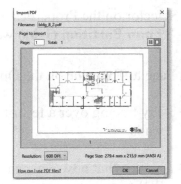

Set the Resolution to **600 DPI**.

Click **OK**.

Left click to place the PDF in the view.

5.
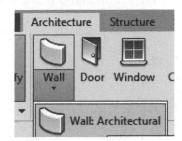

Select the **Wall:Architectural** tool from the Architecture tab on the ribbon.

6.

Use the Type Selector on the Properties panel to select the **Interior – 79 mm Partition (1-hr)** wall.

7. Set the Finish Face to **Exterior** on the Options bar.

8.
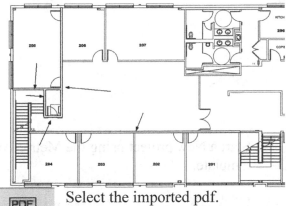

Try tracing over a few of the interior walls.

Notice that you cannot snap to any of the PDF elements.

9. Select the imported pdf.

Toggle **Enable Snaps** on the ribbon so it is ON.

10.
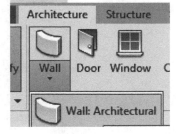

Select the **Wall:Architectural** tool from the Architecture tab on the ribbon.

11.  Use the Type Selector on the Properties panel to select the **Interior – 79 mm Partition (1-hr)** wall.

12. Set the Finish Face to **Exterior** on the Options bar.

13.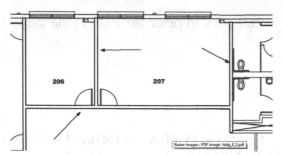

Try tracing over a few of the interior walls.

Notice that now you can snap to the PDF elements to place walls.

14. Save as *ex6-11.rvt*.

Exercise 6-12

Import Image

AUTODESK.
Certified Professional

Drawing Name: wrexham.jpg
Estimated Time: 5 minutes

Scope

❑   Import Image

Solution

1.  Start a New project using the Metric-Architectural template.

Click **OK**.

2. 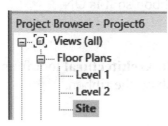 Open the **Site** plan view.

3. 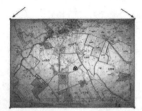 Switch to the Insert tab on the ribbon.

Select **Import Image**.

4. Select *wrexham.jpg*.

Click **Open**.

Left click to place in the view.

5. 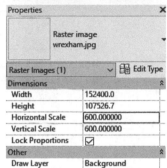 Use the grips at the corners to enlarge the image.

6. On the Properties panel, change the Horizontal Scale to **600**.

Notice that if Lock Proportions is enabled, the Vertical Scale will be the same as the Horizontal Scale.

Properties	
Raster image wrexham.jpg	
Raster Images (1)	Edit Type
Dimensions	
Width	152400.0
Height	107526.7
Horizontal Scale	600.000000
Vertical Scale	600.000000
Lock Proportions	☑
Other	
Draw Layer	Background

7. Save as *ex6-12.rvt*.

Review Warnings

At any time when working on a project, you can review a list of warning messages to find issues that might require review and resolution.

Warnings display in a dialog in the lower-right corner of the interface. When the warning displays, the element or elements that have an error are highlighted in a user-definable color.

Unlike error messages, warning messages do not prohibit the current action. They just inform you of a situation that may not be your design intent. You can choose to correct the situation or ignore it.

The software maintains a list of warning messages that are displayed and ignored while you are working. The Warnings tool lets you view the list at your convenience to determine if the conditions described in the warnings still exist.

If you receive a file from an outside source, it may be worth your time to review any warnings to determine if there are any issues that may require some resolution.

Exercise 6-13
Review Warnings

Drawing Name: review warnings.rvt
Estimated Time: 30 minutes

Scope

- ❏ Review Warnings
- ❏ Export Warnings
- ❏ Areas
- ❏ Stairs
- ❏ Rooms
- ❏ Floors
- ❏ Element ID

Solution

1.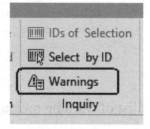

   Switch to the Manage tab on the ribbon.

   Select **Warnings** from the Inquiry panel.

2.

   A list of warnings is displayed.

   Expand **Warning 1**.

3.

Place a check next to the first wall under Warning 1.

Click **Show**.

4.

Click **Close**.

Click **Show** again.

5. All open views that show highlighted elements are already shown. Searching through the closed views to find a good view could take a long time. Continue?

Click **OK**.

6.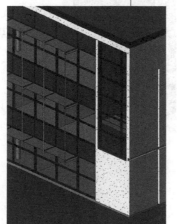

A view will open with the wall highlighted.

Click **Close** to close the dialog.

7. Switch to a 3D view.

8.

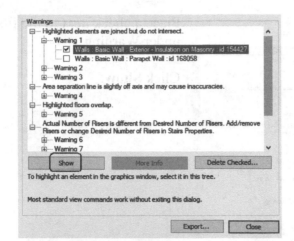

Place a check next to the first wall under Warning 1.

Click **Show**.

9.

The wall highlights.

10.

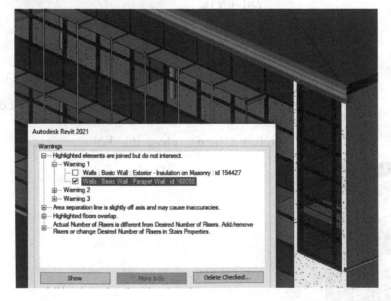

Place a check on the second wall for Warning 1.

The parapet wall highlights.

You can ignore this warning.

11.

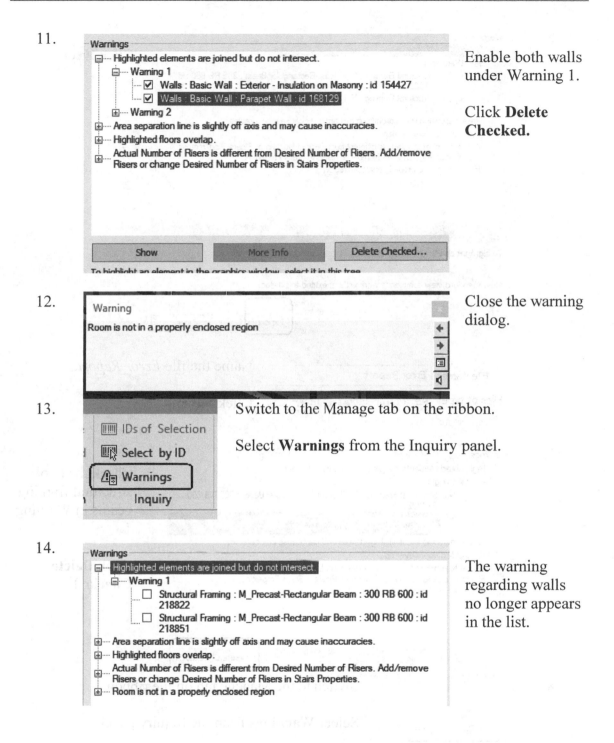

Enable both walls under Warning 1.

Click **Delete Checked.**

12.

Close the warning dialog.

13.

Switch to the Manage tab on the ribbon.

Select **Warnings** from the Inquiry panel.

14.

The warning regarding walls no longer appears in the list.

15.

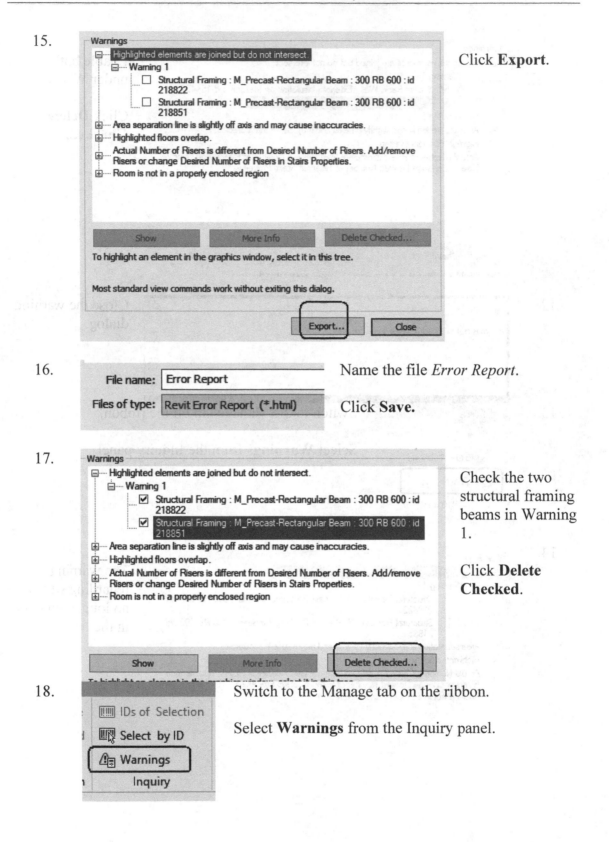

Click **Export**.

16.

Name the file *Error Report*.

Click **Save.**

17.

Check the two structural framing beams in Warning 1.

Click **Delete Checked**.

18.

Switch to the Manage tab on the ribbon.

Select **Warnings** from the Inquiry panel.

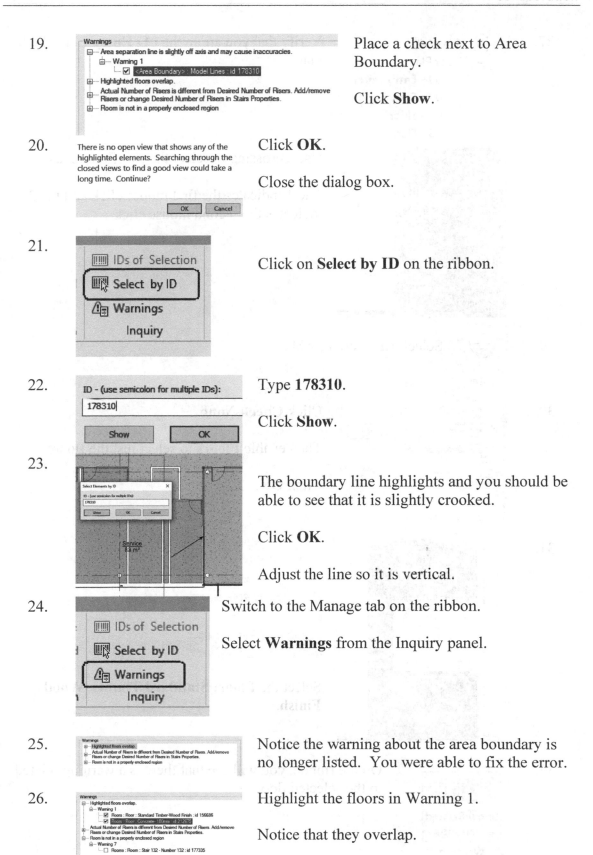

19. Place a check next to Area Boundary.

Click **Show**.

20. Click **OK**.

Close the dialog box.

21. Click on **Select by ID** on the ribbon.

22. Type **178310**.

Click **Show**.

23. The boundary line highlights and you should be able to see that it is slightly crooked.

Click **OK**.

Adjust the line so it is vertical.

24. Switch to the Manage tab on the ribbon.

Select **Warnings** from the Inquiry panel.

25. Notice the warning about the area boundary is no longer listed. You were able to fix the error.

26. Highlight the floors in Warning 1.

Notice that they overlap.

27. 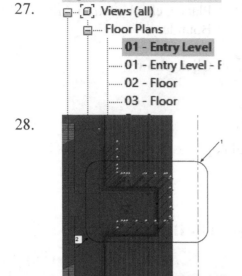 Verify that you are in the **01- Entry Level** floor plan.

28. Use a crossing to select the two floor areas.

The 1 indicates the first mouse click and the 2 indicates the second mouse click.

29. Select **Filter** on the ribbon.

Filter

30. Click **Check None**.

Then enable **Floors** to select just the floors.

Click **OK**.

31.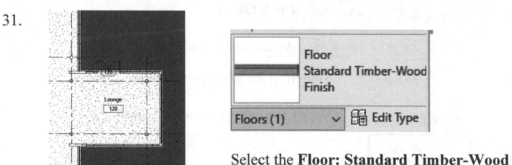

Select the **Floor: Standard Timber-Wood Finish**.

32. On the ribbon, you will see that there is a warning related to the selected floor.

Show Related Warnings

Warning

33.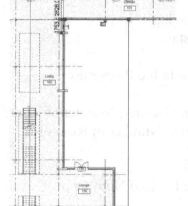

The warning is the warning you already knew about, but it is nice to verify it.

Click **Close** if you opened the Warning dialog.

34.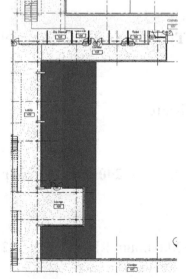

With the **Floor: Standard Timber-Wood Finish** selected, click **Edit Boundary** on the ribbon.

35.

Adjust the boundary so the lines are aligned with the outside edges of the concrete floor.

Green Check to exit the editing mode.

36.

Reselect the floor.

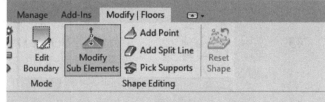

Check to see if the Warning icon appears on the ribbon.

If it does, edit the boundary again and adjust so it no longer overlaps the concrete floor.

37.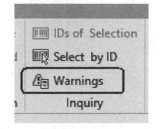

Switch to the Manage tab on the ribbon.

Select **Warnings** from the Inquiry panel.

38.

Expand Warnings 1 through 5.

Can you tell if these warnings apply to only one element or how many elements need to be modified?

39.

Place a check next to the Stairs under Warning 1.

Click **Show**.

40.

Close the dialog.
Select the stairs.

Dimensions	
Width	1625.0
Desired Number of Risers	26
Actual Number of Risers	25
Actual Riser Height	146.2
Actual Tread Depth	300.0

Scroll down in the Properties panel.

Note that the Desired Number of Risers is 26, but the Actual Number of Risers is 25.

41.

Desired Number of Risers is too small. Computed Actual Riser Height is greater than Maximum Riser Height allowed by type.

Change the Desired Number of Risers to 25.

This error will appear.

Click **Close**.

42.

Stair
150mm max riser 300mm tread

Click **Edit Type** on the Properties panel.

43.

Family: System Family: Stair
Type: 150mm max riser 300mm tread

Select **Duplicate**.

44.

Name: 200mm max riser 300mm tread

Change the Name to **200mm max riser 300 mm tread.**

Click **OK.**

45.

Type Parameters	
Parameter	
Calculation Rules	
Calculation Rules	
Minimum Tread Depth	300.0
Maximum Riser Height	200.0

Change the Maximum Riser Height to **200mm**.

Click **OK.**

46.

Properties	
Stair 200mm max riser 300mm tread	
Stairs (1)	
Show Up arrow Stairs (1) s	☐
Dimensions	
Width	1625.0
Desired Number of Risers	25
Actual Number of Risers	25
Actual Riser Height	152.0
Actual Tread Depth	300.0

The Stair now shows the Desired Number of Risers equal to the Actual Number of Risers.

The Actual Riser Height is 152 mm which is about 6 inches high, well within code.

47.

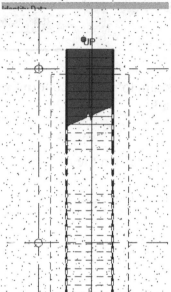

Select the stairs above the stairs you just modified.

This stair has the same error – the desired and actual number of risers do not match.

Use the Type Selector to change the Stair to the new type.

Properties	
Stair 150mm max riser 300mm tread	
Stairs (1)	
Up label	☑
Up arrow	☑
Down label	☑
Down arrow	☑
Show Up arrow in all views	☐
Dimensions	
Width	1625.0
Desired Number of Risers	26
Actual Number of Risers	25
Actual Riser Height	146.2
Actual Tread Depth	300.0
Identity Data	

48.

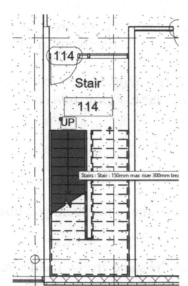

Locate the stair in Room 114.

Right click and select **Select All Instances→In Entire Project.**

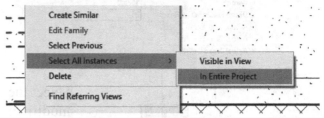

Change the stairs to use the **200mm max riser 300 mm tread.**

49.

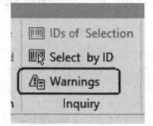

Switch to the Manage tab on the ribbon.

Select **Warnings** from the Inquiry panel.

50.

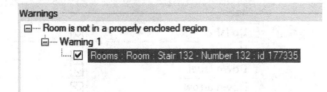

Expand Warning 1.

Click **Show**.

Close the dialog.

51.

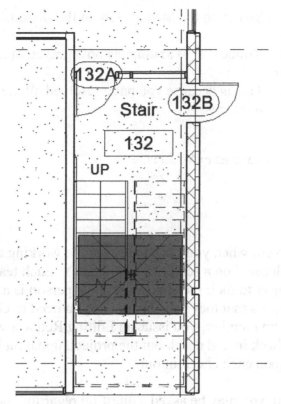

The room is located in Stair 132, but there is no wall on the south side, so the room is not fully enclosed.

52.

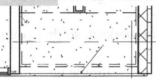

Switch to the Architecture tab on the ribbon.

Select **Room Separator** on the Room & Area panel.

53.

Draw a line between the two vertical lines to enclose the room.

54.

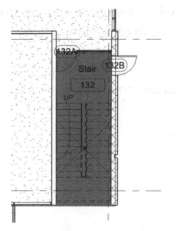

Use a crossing and a filter to select the room again.

The room size has adjusted using the walls and the room separator.

55. 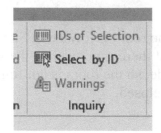 Switch to the Manage tab on the ribbon.

Notice that **Warnings** is now grayed out.

This is because you have resolved all the warnings.

56. Save as *ex6-13.rvt*.

Worksets

Worksets are used in a team environment when you have many people working on the same project file. The project file is located on a server (a central file). Each team member downloads a copy of the project to their local machine. The person is assigned a workset consisting of building elements that they can change. If you need to change an element that belongs to another team member, you issue an Editing Request which can be granted or denied. Workers check in and check out the project, updating both the local and server versions of the file upon each check in/out.

In the Professional Certification exam, you may be asked a question regarding how to control the display of worksets or how to identify elements in a workset.

Other Hints:

- Name any sheets you create. That way you can distinguish between your sheet and other sheets.
- Create a view that is specific to your changes or workset, so that you have an area where you can keep track of your work.
- Create one view for the existing phase and one view for the new construction phase. That way you can see what has changed. Name each view appropriately.
- Use View Properties to control the phase applied to each view.
- Check Editing Requests often.
- Duplicate any families you need to modify, rename, and redefine. If you modify an existing family, it may cause problems with someone else's workset.

Exercise 6-14

Worksets

Drawing Name: **worksets.rvt**
Estimated Time to Completion: 15 Minutes

Scope

Use of Worksets
Workset Visibility

Solution

1.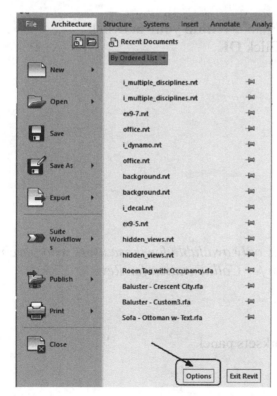

Before you can use Worksets, you need to set Revit to use your name.

Close any open projects.

Go to **Options**.

2. 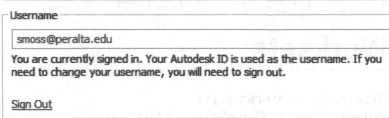 Select the **General** tab. Enter your name in the Username text field. Click **OK**.

If you are using the Autodesk account, you will see the user name assigned to that account. To use a different name, sign out from the Autodesk account.

3. Locate the worksets project. Click **Open**.

4. Collaborate   Select the **Collaborate** tab on the ribbon.

5. 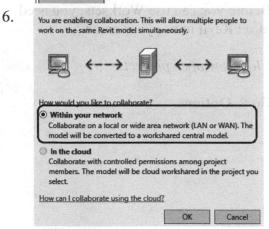 Select the **Collaborate** tool.
   There will be a slight pause while Revit checks to see if you are connected to the Internet.

6. Enable **Within your network**.
   Click **OK**.

Collaborate in Cloud is only available for those users who purchase a separate subscription for Collaboration for Revit.

7. Select **Worksets** on the Worksets panel.

8.

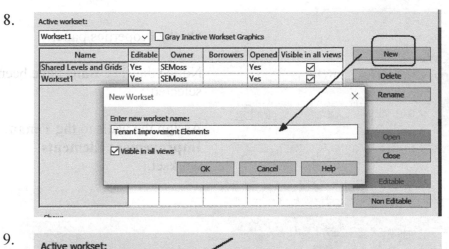

Create a new worksets by clicking **New.**

Type **Tenant Improvement Elements** for the name.

Click **OK**.

9.

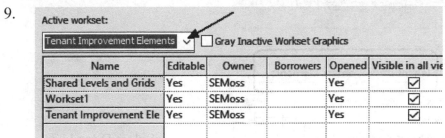

Use the drop-down to set the Active workset as **Tenant Improvement Elements.**

Click **OK.**

10.

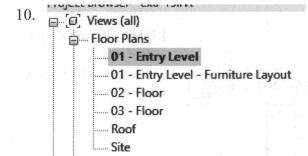

Activate the **01- Entry Level** floor plan.

11.

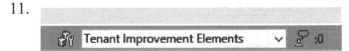

Notice that the **Tenant Improvement Elements** is displayed as the active workset.

12.

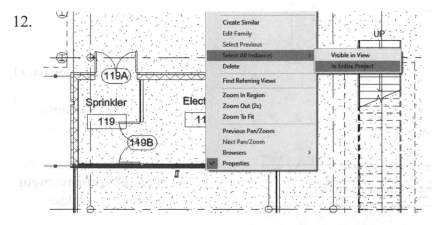

Select one of the interior walls.

Right click and select **Select All Instances→In Entire Project**.

13.

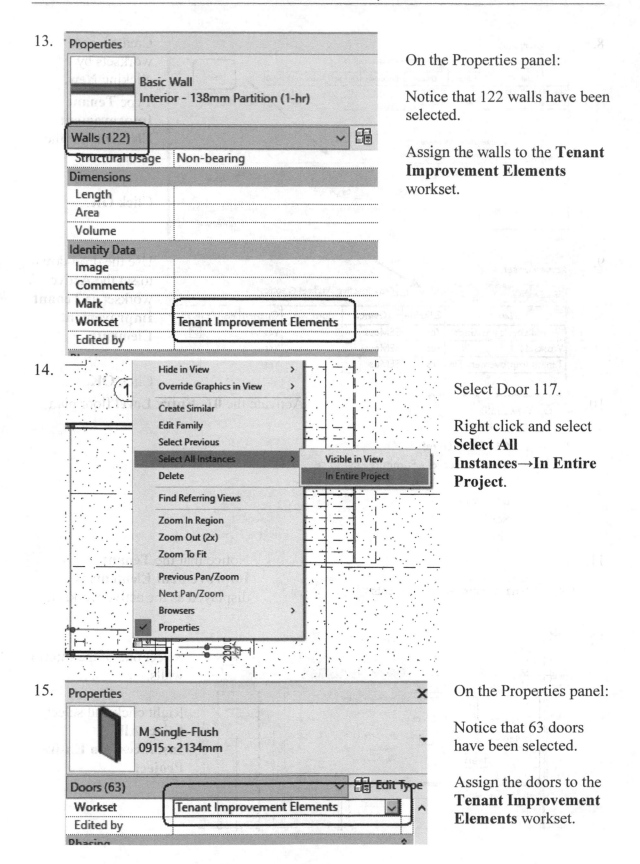

On the Properties panel:

Notice that 122 walls have been selected.

Assign the walls to the **Tenant Improvement Elements** workset.

14.

Select Door 117.

Right click and select **Select All Instances→In Entire Project**.

15.

On the Properties panel:

Notice that 63 doors have been selected.

Assign the doors to the **Tenant Improvement Elements** workset.

16.

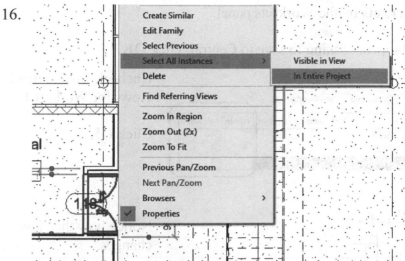

Select Door 118.

Right click and select **Select All Instances→In Entire Project**.

17.

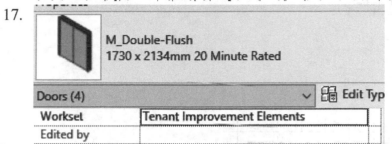

On the Properties panel:

Notice that 4 doors have been selected.

Assign the doors to the **Tenant Improvement Elements** workset.

18.

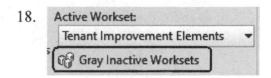

On the Collaborate tab on the ribbon:

Enable **Gray Inactive Worksets**.

19.

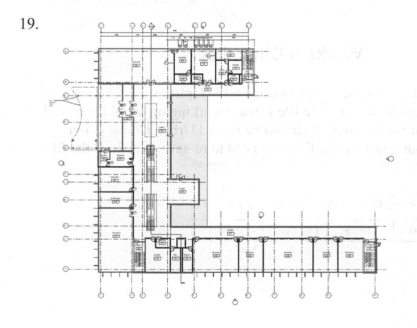

Notice how the display changes to highlight only those elements assigned to the active workset.

Disable **Gray Inactive Worksets** to restore the view.

20. Select **Worksets** on the Worksets panel.

    Worksets If a dialog appears asking to save to Central, click **OK**.

21.

Active workset:					
Tenant Improvement Elements ∨ ☐ Gray Inactive Workset Graphics					
Name	Editable	Owner	Borrowers	Opened	Visible in all views
Shared Levels and Grids	Yes	SEMoss		Yes	☑
Tenant Improvement Ele	Yes	SEMoss		Yes	☑
Workset1	Yes	SEMoss		Yes	☐

Uncheck Visible in all views for Workset 1.

Click **OK**.

22.

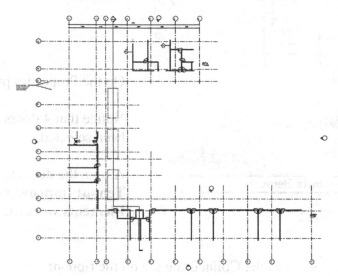

Notice how the view changes.

23. Save as *ex6-14.rvt*.

Workset Checklist

1. Before you open any files, make sure your name is set in Options.

2. The main file is stored on the server. The file you are working on should be saved locally to your flash drive or your folder. It should be named Urban House [Your Name]. If you don't see your name on the file, you need to re-save or re-load the file.

Active workset:					
Workset1 (Not Editable) ▼ ☐ Gray Inactive Workset Graphics					
Name	Editable	Owner	Borrowers	Opened	Vis
Shared Levels and Grids	No			Yes	
Workset 2	Yes	Elise		Yes	
Workset1	No			Yes	

3. Go to Work sets and verify that your name is next to the Work set you are assigned.

4. Verify that the Active work set is the work set you are assigned.
5. Re-load the latest from Central so you can see all the updates done by other users.
6. When you are done working, save to Central – so other users can see your changes.
7. When you close, relinquish your work sets, so others can keep working.

The Worksharing Display Options allow you to color code elements in a view to distinguish between which elements different team members own as well as identifying which worksets different elements have been assigned.

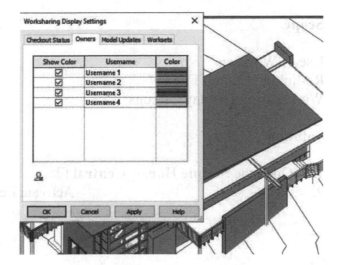

Exercise 6-15
Worksets Display

Drawing Name: **Simple House_Central.rvt**
Estimated Time to Completion: 90 Minutes

Scope

Use of Worksets
Re-linking Central File
Worksharing Display Options

Solution

1. Open the **Simple House_Central** file.
2. VIEWS (all)   Activate Level 1.
    Floor Plans
        Level 1
        Level 1 - Room Legend
        Level 2
        Site

3.   Select **Worksharing Display Settings**
                                          on the View Control bar.

4.   On the Checkout Status tab:

                                          Enable the **Show Color** display for all
                                          the options.

                                          Click **OK**.

5.

Enable display of **Worksets**.

6.

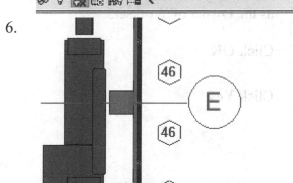

What color is Column Grid E?

It displays as Green which indicates it is owned by the Administrator.

7.

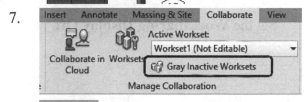

Switch to the Collaborate tab on the ribbon.

Enable **Gray Inactive Worksets**.

8.

Select the **Worksets** tool from the Collaborate tab on the ribbon.

9.

The central model D:\Revit 2021 Certification Guide\Revit 2021 Certification Guide Exercises\Simple House_Central.rvt cannot be found, perhaps due to a lost network connection.

Check out worksets to edit currently non-editable elements. If other users have edited these elements, all changes you made in this file will be lost when you reconnect to the central model.

Close

If you see this error, click **Close**.

10.

Active workset:					
Workset1 (Not Editable)	☑ Gray Inactive Workset Graphics				
Name	Editable	Owner	Borrowers	Opened	Visible in all views
Workset1	No			Yes	☑
Workset2	No			Yes	☑
Workset3	No			Yes	☑
Workset4	No			Yes	☑

Set all the worksets Editable column to **No**.

This relinquishes ownership of those worksets.

11. 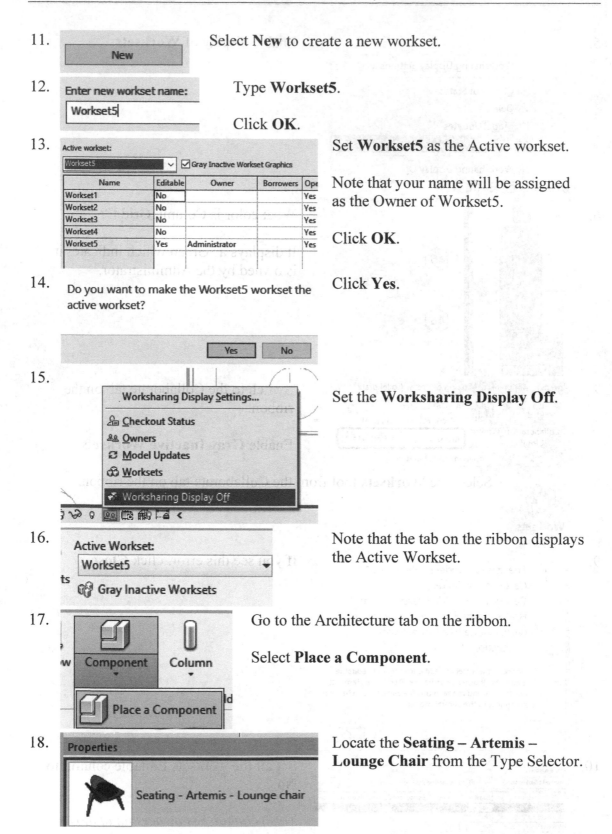 Select **New** to create a new workset.

12. **Enter new workset name:** Type **Workset5**.

Workset5

Click **OK**.

13. Set **Workset5** as the Active workset.

Note that your name will be assigned as the Owner of Workset5.

Click **OK**.

14. Do you want to make the Workset5 workset the active workset? Click **Yes**.

15. **Worksharing Display Settings...** Set the **Worksharing Display Off**.

Checkout Status
Owners
Model Updates
Worksets
Worksharing Display Off

16. **Active Workset:** Note that the tab on the ribbon displays the Active Workset.

Workset5

Gray Inactive Worksets

17. Go to the Architecture tab on the ribbon.

Select **Place a Component**.

Component   Column

Place a Component

18. **Properties** Locate the **Seating – Artemis – Lounge Chair** from the Type Selector.

Seating - Artemis - Lounge chair

19.

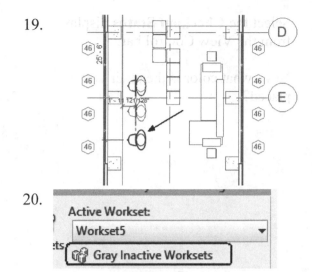

Place a chair below the other two chairs.

Use the SPACE bar to rotate the chair prior to placing.

Note that the chair you placed is not grayed out.

20.

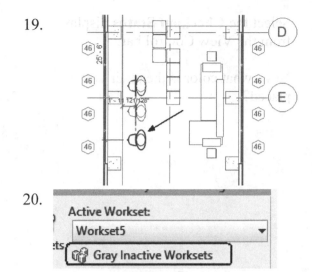

Select the **Gray Inactive Worksets** toggle on the Collaborate tab on the ribbon.

Note that the display changes so nothing is grayed out.

21.

Select the **Worksets** display from the View Control bar.

22.

What color is the chair you placed?

23.

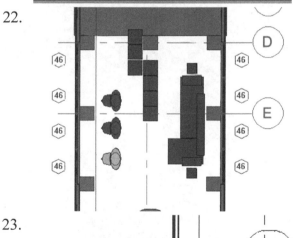

Select the **Owners** display from the View Control bar.

Now, what color is the chair you placed?

24.

> Worksharing Display Settings...
>
> 👤 Checkout Status
> 👥 Owners
> 🔁 Model Updates
> 🔁 Worksets
>
> 🔁 Worksharing Display Off

Select the **Checkout Status** display from the View Control bar.

Now, what color is the chair you placed?

25. Close the file without saving.

Certified Professional Practice Exam

1. Worksharing allows team members to:

    A. Share families from different projects
    B. Share views from different projects
    C. Work on the same parts of a project simultaneously
    D. Work on different parts of the same project

2. Set Phase Filters for a view in the:

    A. Properties pane
    B. Design Options
    C. Tab on the ribbon
    D. Project Browser

3. A rectangular column family comes with three types: 18″ x 18″, 18″ x 24″, and 24″ x 24″. To add a 6″ x 6″ type, you:

    A. Duplicate, rename and modify the instance parameters
    B. Duplicate, rename and modify the type parameters
    C. Rename and modify the type parameters
    D. Duplicate, rename the family and add a type

4. The Coordination Review tool is used when you use:
    A. Worksets
    B. Linked Files
    C. Interference Checking
    D. Phase

5. The Worksets option under the worksharing display menu will show:

    A. The color that corresponds with each workset
    B. The color that corresponds with the active workset
    C. The color that corresponds with the owner of the active workset
    D. The color of all worksets that can be edited (checked out)

6. True/False
    You can ignore review warnings, but you cannot ignore an error.

7. The Coordination Review is useful for:

    A. Checking what monitored elements were changed in the host file
    B. Aligning the origin of one project with a linked project
    C. Performing multiple copy operations from a linked project
    D. Copying a large number of elements from one project to another

8. When you link to a CAD drawing:

   A. You bring in an image of the CAD file which cannot be updated
   B. You automatically import any external references used by that CAD file.
   C. If the CAD file is modified, you will see any changes when you reload the file.
   D. The CAD file is added to the Revit project as decal.

9. You are using worksets. When you make changes to the local file, these changes:

   A. Cannot be saved to the Central Model.
   B. Can only be saved to the Central Model if the changes are minor.
   C. Are reflected in the Central Model only after synchronizing.
   D. Are automatically reflected in the Central Model.

10. When reviewing errors, why isn't using the Remove Constraints button the best method to resolve the error?

   A. It might not actually remove the constraints
   B. It may add more constraints
   C. It's too easy a solution
   D. It may remove elements that are required in the design

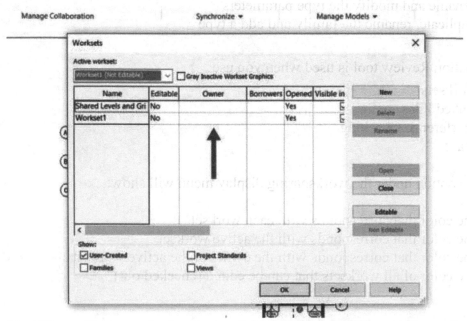

11. What will appear in the Owner field for Workset1 if you change the Editable field to Yes?

   A. The User name you set in the Options will appear in the Owner field
   B. The original creator of the workset will be assigned as the Owner
   C. Workset1 will not be assigned an Owner, but it will still be editable
   D. The User you select from the drop-down list will be assigned the owner

12. I choose Import CAD to bring in an AutoCAD drawing file with the floor plan. My vendor just emailed me that he made some changes to the floor plan. What now?

    A. An asterisk will appear next to the drawing file name indicating there are unsaved changes
    B. The drawing will automatically update
    C. A warning will pop up the next time I reopen the Revit project advising me to reload the drawing
    D. The changes are not reflected in my Revit project and I do not see any warnings or indications that the file may have changed.

13. When creating a local version of a centralized model, what happens if you select the Detach from Central option when opening the file?

    A. You will have a read-only version of the model and can't make changes but can review.
    B. Changes you make will automatically synchronize with the Central Model
    C. Changes you make to the local model won't be synchronized and the local file will now be independent of the Central model.
    D. Your username will not be assigned to any worksets.

14. You import a PDF into your Revit project, but you are unable to snap to any elements in the PDF. You should:

    A. Enable Snap on the ribbon
    B. Convert the PDF to Revit elements
    C. Reload the PDF
    D. Just do the best you can tracing over the PDF

15. To review any warnings within a project:

    A. Go to the Manage tab on the ribbon and select Warnings.
    B. Go to the Collaborate tab on the ribbon and select Warnings.
    C. Select an element, right click and select Warnings
    D. Select an element and click on Show Related Warnings on the ribbon

16. Which icon should be used to manage the display settings for worksets?

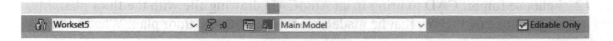

17. Editable Only is enabled on the status bar. This means:

    A. Only elements in the Main Model design option can be edited
    B. Only elements in the host file can be edited
    C. Only elements in the active workset can be edited
    D. Only elements in the worksets with editing enabled can be selected or edited.

Answers
    1) D; 2) A; 3) D; 4) C; 5) B; 6) T; 7) A; 8) C; 9) C; 10) D; 11) A; 12) D; 13) C; 14) A; 15) A; 16) C; 17) D

Revit Utilities

This lesson addresses the following Professional certification exam questions:

- Audit
- Purge
- Save sheet as dwg
- Transfer project standards
- Etransmit

Audit

Use the Audit function periodically to maintain the health of a Revit model, when preparing to upgrade the software, or as needed to locate and correct issues.

Auditing an Autodesk Revit file helps you maintain the integrity of large files and files that more than one user works on. The auditing process checks your file integrity to determine that no corruption has occurred within the file's element structure and automatically fixes any minor file errors that it encounters. You can perform an audit on your Revit file when you open it in the application.

The Audit function scans, detects, and fixes corrupt elements in the model. It does not provide feedback on which elements are fixed.

Auditing a model can be time-consuming, so be prepared to wait while the process completes.

As a best practice, audit the model weekly. If the model is changing rapidly, audit it more often.

In addition to auditing models, you can also use this function to audit families and templates.

Exercise 7-1

Audit a File

Drawing Name: **audit.rvt**
Estimated Time to Completion: 5 Minutes

Scope
Audit a Revit building project

Solution

1.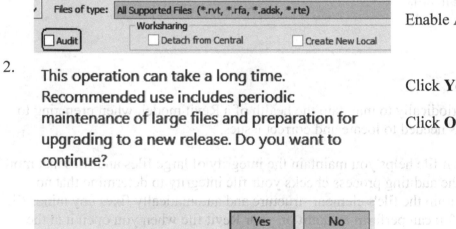

    Locate the audit.rvt file.

    Enable **Audit**.

2. This operation can take a long time. Recommended use includes periodic maintenance of large files and preparation for upgrading to a new release. Do you want to continue?

    Click **Yes**.

    Click **Open.**

    | Yes | No |

3. Save the file and close.

Purge

Purging a Revit model assists in removing unused families, views, and objects from a project. It is recommended that the model should be purged after every submittal and milestone to remove any remaining elements that have accumulated in the project. Revit's automated purging method only removes certain elements within your project; therefore, you still need to go through your models and manually remove any unwanted area schemes, views, groups, and design options. You should also replace in-place families with regular component families to further reduce the file size.

When you consider using **Purge Unused** it is important to understand how this works, mainly because in most cases should you require to remove any unwanted families, objects and views, you would need to employ the 3 step Purge process. The reason for this is when first using *Purge Unused* you can only see the items that are not being currently used in the project. This means that any families that are listed may be using materials that are in the project. So when you purge out a family, its materials now might be unused in the project, and when you next launch the Purge Unused dialog you may see additional materials that can be purged from the project. As a recommendation it is also necessary to select the *Check None* button as your first step when using *Purge Unused*, then look for and select what you would like to purge. This is important because by default everything that can be purged is pre-selected. This means if you have an element in your project template, but you haven't used it yet in the project, it will be removed and you'll have to re-create it or re-load it.

I try to discourage my students from using the Purge tool because they invariably find themselves needing families they have purged and then they have to spend a great deal of time trying to locate them and reload them.

Always remember to leave space for file growth because files that become large in the early stages will only grow larger in the subsequent design development stages. For example, if you have 16GB of RAM an overall guideline would be that anything below 200MB is safe. If the file climbs above 300MB in schematic design or design development stages, you should consider restructuring the model. Should it get above 500MB at any stage, then employing the *Purge Unused* command is strongly suggested.

One way to reduce file size is to use linked files. You can then unload the linked files and reload them when you need to see how elements interact.

Exercise 7-2

Purge a File

Drawing Name: **purge.rvt**
Estimated Time to Completion: 5 Minutes

Scope
Purge a Revit building project

Solution

1. Switch to the Manage tab on the ribbon.

   Select **Purge Unused** from the Settings panel.

2.

Expand the list.

Notice there is an unused Legend.

This means the legend has not been placed on a sheet, but you may still want to use it.

Browse through the list.

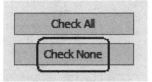

Click **Check None**.

That way you don't accidentally delete an element you may need later.

3.

Locate the **Linear Dimension Style**.

Select the two dimension styles listed.

Click **OK**.

4. Save as *ex7-2.rvt*.

Save Sheet As DWG

Revit provides the option to export sheets as dwg files. You can specify layer names, fonts, and colors to be used for the export.

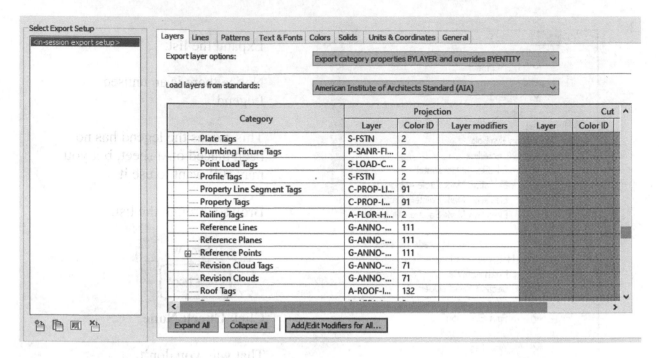

For Export layer options, specify how Revit elements with view-specific graphic overrides will be mapped to CAD layers.

(Graphic properties of Revit elements, such as color, line weight, and line style, are defined in the <u>Object Styles</u> of the categories to which the elements belong, but these definitions can be <u>overridden</u> for a selected element in a specific view.)

- **Export category properties BYLAYER and overrides BYENTITY**. A Revit element with view-specific graphic overrides will retain those overrides in the CAD application but will reside on the same CAD layer as other entities in the same Revit category.

- **Export all properties BYLAYER, but do not export overrides**. View-specific graphic overrides will be ignored in the CAD application. Any exported Revit element will reside on the same CAD layer as other entities in the same Revit category. By forcing all entities to display the visual properties defined by their layer, this option results in a lower number of layers and provides by-layer control over the exported DWG/DWF file.

- **Export all properties BYLAYER, and create new layers for overrides**. A Revit element with view-specific graphics will be placed on its own CAD layer. This option provides by-layer control over the exported DWG/DXF file and preserves graphical intent. However, it increases the number of layers in the exported DWG file.

Note: If you are exporting a view that contains a linked project and the RVT Link Display Settings dialog (Basics tab) for the link is set to By Host View, then the link is treated as an override. To ensure that colors and other graphic display settings are preserved in the exported file, select "Export all properties BYLAYER, but do not export overrides."

You can also opt to hide reference planes and scope boxes.

⋯⋯ Property Line Segment Tags	C-PROP-LINE-IDEN	91	
⋯⋯ Property Tags	C-PROP-IDEN	91	
⋯⋯ Railing Tags	A-FLOR-HRAL-IDEN	2	
⋯⋯ Reference Lines	G-ANNO-SYMB	111	
⋯⋯ Reference Planes	G-ANNO-SYMB	111	Add/Edit...
⊞⋯ Reference Points	G-ANNO-SYMB	111	
⋯⋯ Revision Cloud Tags	G-ANNO-REVC-IDEN	71	
⋯⋯ Revision Clouds	G-ANNO-REVC	71	
⋯⋯ Roof Tags	A-ROOF-IDEN	132	

Reference Planes are placed on the G-ANNO-SYMB layer by default. You may be asked how Revit treats reference planes, lines, and fonts during the export process.

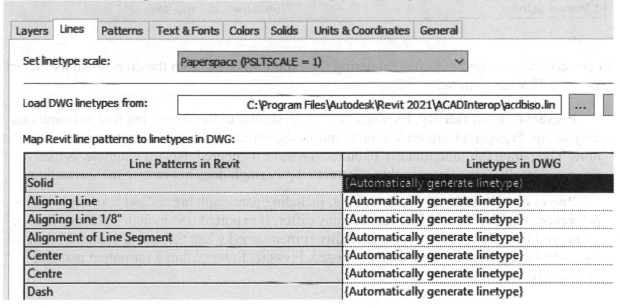

When you export a 2D view to a DWG file, lines are treated as follows:

- If two lines overlap in the drawing, the thicker of the lines is retained. The thinner line is shortened or removed.

- If a thick line is shorter than a thin line and its start point and endpoint fall within the thin line, no action occurs.

- If two collinear lines with the same visual parameters overlap, they are merged into one.

- When walls become lines in the DWG file, no short collinear lines are produced.

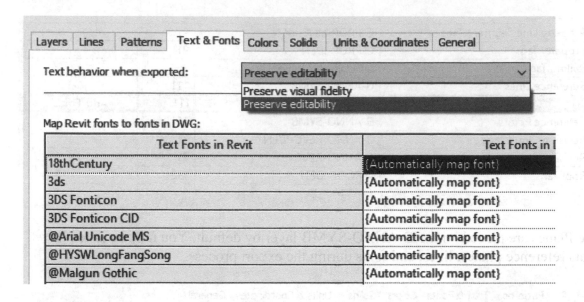

You can control how text is formatted during the export process. From the drop-down list, select one of the following options:

- **Preserve visual fidelity.** Exported text looks similar to Revit text, but font substitutions may occur. Paragraphs are broken up to mimic the source text rather than wrapped in the text note. The paragraph functionality of bulleted lists or numbered lists for example, is likely lost upon export. Pressing Enter within a formatted paragraph does not restore the formatting.

- **Preserve editability.** Text formatting, including paragraph breaks and text wrapping, is preserved. Positioning and line lengths may differ. If exported text includes bulleted lists or numbered lists, that paragraph functionality is maintained when the text is edited. Multilevel lists will likely need to be manually adjusted. Pressing Enter within a formatted paragraph restores the formatting.

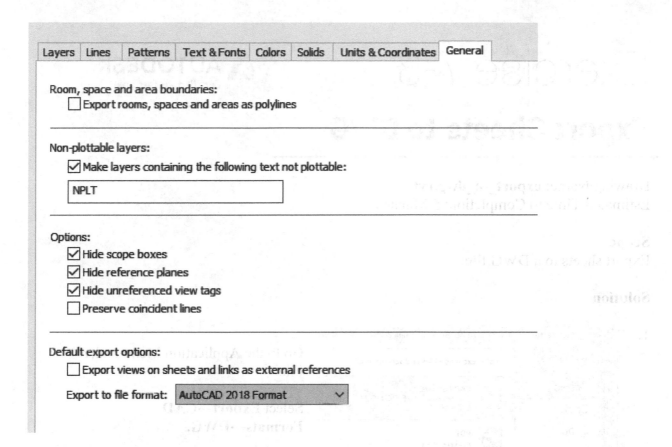

On the General tab of the Export settings, you can specify whether or not views on sheets should be treated as external references. If this is enabled individual files are created for each view. You can also specify which version of AutoCAD to be used for the export.

Exercise 7-3

Export Sheets to DWG

Drawing Name: **export_to_dwg.rvt**
Estimated Time to Completion: 5 Minutes

Scope
Export sheets to a DWG file.

Solution

1.

Go to the Application Menu under File.

Select **Export→CAD Formats→DWG.**

2.

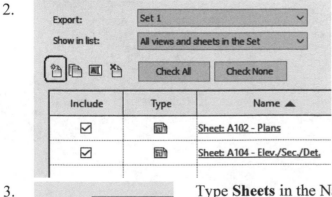

Select **Set 1**.
Select **New Set**.

3.

Name: Sheets|

Type **Sheets** in the Name field.

Click **OK**.

4.

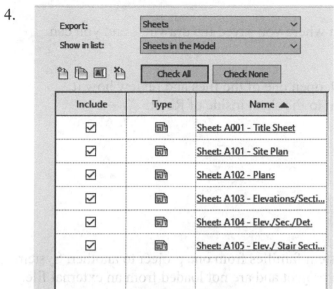

Select **Sheets in the Model** from the drop-down list.

Click **Check All**.

This includes all sheets in the drawing export.

Click **Next.**

5.

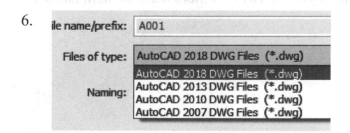

Set the **Naming to Automatic – Short.**

This will use each sheet name as the file name.

If you select Automatic Long, Revit will use the Revit project name as a prefix or you can specify your preferred prefix to add to the file names.

Disable **Export views on sheets and links as external references**.

6.

Under Files of type: you can specify which version of AutoCAD to use.

Select **AutoCAD 2018**.

Click **OK.**

7.

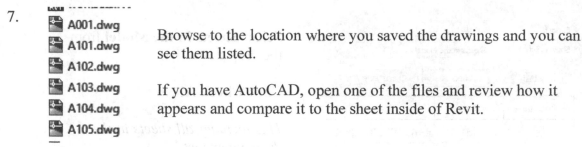

Browse to the location where you saved the drawings and you can see them listed.

If you have AutoCAD, open one of the files and review how it appears and compare it to the sheet inside of Revit.

8. Save as *ex7-3.rvt*.

Transfer Project Standards

Transfer Project Standards is used to copy system families from one project to another. System families are elements that are defined inside a project and are not loaded from an external file. System families include view templates, walls, conduits, and wires.

You need to have the project open that you want to import into as well as the project you want to transfer from. If the project you are transferring from has a linked file, you can also import families from the linked file without opening that file.

Items which can be copied from one project to another include:

- Family types (including system families, but not loaded families)
- Line weights, materials, view templates, and object styles
- Mechanical settings, piping, and electrical settings
- Annotation styles, color fill schemes, and fill patterns
- Print settings

You may get a prompt during the transfer process alerting that an item being transferred already exists in the new project. You may opt to overwrite, ignore or cancel for the existing elements. You do not get to pick and choose which elements you overwrite if more than one item has been selected to be transferred.

Exercise 7-4

Transfer Project Standards

Drawing Name: *export standards.rvt*
Estimated Time: 10 minutes

Scope

Transfer Project Standards

Solution

1. 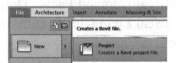 Open the *export standards.rvt file.*
This is the file we will import standards from.

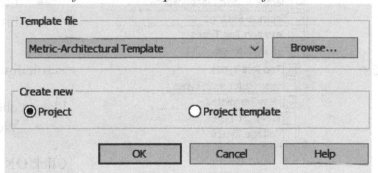

Start a New Project using the Metric-Architectural default template.

2. 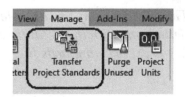 Select the Manage tab on the ribbon.

Select **Transfer Project Standards** from the Settings panel on the ribbon.

3.

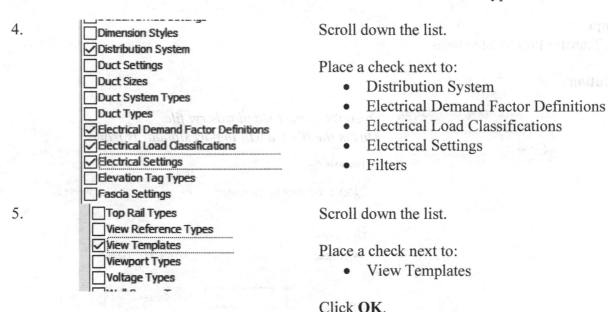

Select *export standards.rvt* to copy from.

This file must be open to be selected.

Click **Check None**.

Place a check next to:
- Conduit Settings
- Conduit Sizes
- Conduit Standard Types
- Conduit Types

4.

Scroll down the list.

Place a check next to:
- Distribution System
- Electrical Demand Factor Definitions
- Electrical Load Classifications
- Electrical Settings
- Filters

5.

Scroll down the list.

Place a check next to:
- View Templates

Click **OK**.

6.

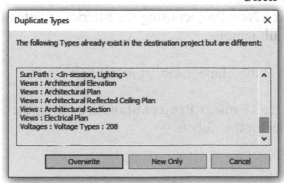

Click **Overwrite**.

This replaces any existing element definitions with the definitions used in the export standards file.

7.

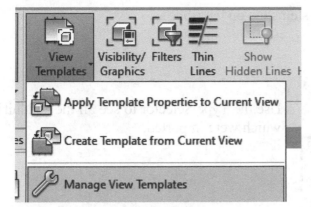

Switch to the View tab on the ribbon.

Select **Manage View Templates**.

8.

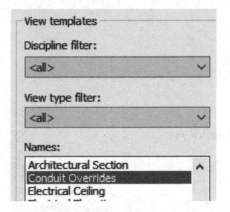

You imported a view template called Conduit Overrides.

Click **OK** to close the dialog box.

9.

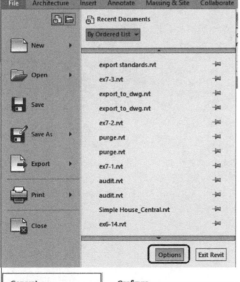

Select **File→Options**.

10.

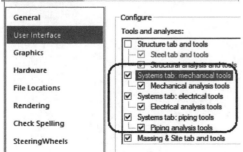

Highlight the User Interface category.

Enable the **Systems tab** for mechanical, electrical and piping.

Click **OK**.

11.   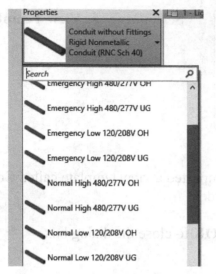   Switch to the Systems tab on the ribbon.
Conduit   Select **Conduit**.

12.   Use the Type Selector to see all the conduit types
which were imported.

Properties   × 1 - Lig

Conduit without Fittings
Rigid Nonmetallic
Conduit (RNC Sch 40)

Search

Emergency High 480/277V OH

Emergency High 480/277V UG

Emergency Low 120/208V OH

Emergency Low 120/208V UG

Normal High 480/277V OH

Normal High 480/277V UG

Normal Low 120/208V OH

Normal Low 120/208V UG

13.   Save as *ex7-4.rvt*.

eTransmit

eTransmit allows you to copy a Revit project along with any linked/dependent files to a single folder. This is also useful if you want to create snapshots of your project at different stages of construction.

You can elect to:
- Include related dependent files such as linked models and DWF markups.
- Include supporting files such as documents or spreadsheets.
- Upgrade the Revit (.rvt) model and linked models to the current release.
- Disable worksets.
- Remove unused families, materials, and other objects from the Revit models to reduce file size.
- Delete sheets, and specific view types so that the models do not contain unnecessary data.
- Include only the views that are placed on sheets.

Shared parameter files, external font files, and lookup tables are not included in the eTransmit package.

Exercise 7-5

A AUTODESK.
Certified Professional

Transfer a Model
Transmit a Model

Drawing Name: etransmit.rvt
Estimated Time: 25 minutes

Scope
  eTransmit

Solution

1.

    Verify all files are closed.
    Select the **Add-Ins** tab on the ribbon.
    Select **Transmit a model**.

2.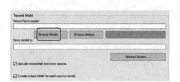

    Click on **Browse Model...**

3.	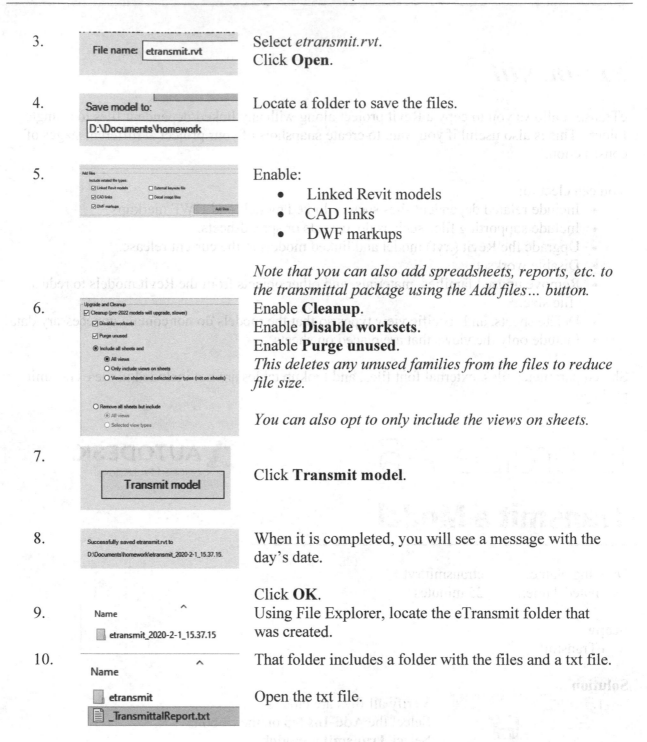File name: etransmit.rvt	Select *etransmit.rvt*. Click **Open**.
4.	Save model to: D:\Documents\homework	Locate a folder to save the files.
5.	Add Files Include related file types: ☑ Linked Revit models ☐ External keynote file ☑ CAD links ☐ Decal image files ☑ DWF markups Add files...	Enable: • Linked Revit models • CAD links • DWF markups *Note that you can also add spreadsheets, reports, etc. to the transmittal package using the Add files... button.*

6.	Upgrade and Cleanup
	☑ Cleanup (pre-2022 models will upgrade, slower)
	☑ Disable worksets
	☑ Purge unused
	◉ Include all sheets and
	  ◉ All views
	  ○ Only include views on sheets
	  ○ Views on sheets and selected view types (not on sheets)
	
	○ Remove all sheets but include
	  ◉ All views
	  ○ Selected view types
		Enable **Cleanup**.
		Enable **Disable worksets**.
		Enable **Purge unused**.
		This deletes any unused families from the files to reduce file size.
		
		You can also opt to only include the views on sheets.

7.	Transmit model	Click **Transmit model**.
8.	Successfully saved etransmit.rvt to D:\Documents\homework\etransmit_2020-2-1_15.37.15.	When it is completed, you will see a message with the day's date. Click **OK**.
9.	Name ^ etransmit_2020-2-1_15.37.15	Using File Explorer, locate the eTransmit folder that was created.

10.	Name ^
	  etransmit
	  _TransmittalReport.txt
		That folder includes a folder with the files and a txt file.
		
		Open the txt file.

11.  Note the report lists which files were included as well as which files were excluded.

Certified Professional Practice Exam

1. I want to use Transfer Project Standards to copy a dimension family to another project. What sub-families can be included? (Select Two)

   A. Text Types
   B. Arrowheads
   C. Lines
   D. Filled Regions

2. True/False I can turn off the visibility of reference planes if I save a Revit sheet to a dwg file.

3. True/False View templates and filters should be transferred together in order to maintain their relationships.

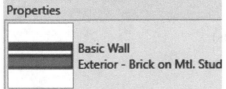

Family:	Basic Wall
Type:	Exterior - Brick on Mtl. Stud
Total thickness:	350.0
Resistance (R):	9.4859 (m²·K)/W
Thermal Mass:	13.41 kJ/K

Layers

		Function	Material	EXTERIOR S
1	Finish 1 [4]		Brick, Common	90.0
2	Thermal/Air Layer [3]		Air	76.0
3	Membrane Layer		Air Infiltration Barrier	0.0
4	Substrate [2]		Plywood, Sheathing	19.0
5	Core Boundary		Layers Above Wrap	0.0
6	Structure [1]		Metal Stud Layer	152.0

4. You have a project that uses a Basic Wall Exterior – Brick on Mtl Stud. The wall family uses the Brick, Common material.

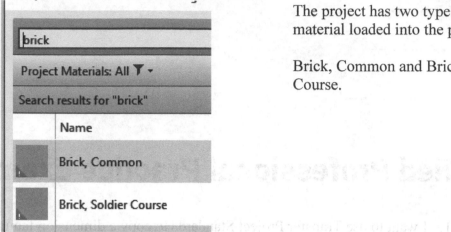

The project has two types of Brick material loaded into the project:

Brick, Common and Brick, Soldier Course.

You use Transfer Project Standards and select *only* Wall Types to copy the wall families over to a new project. Choose what happens:

A. Only the Brick, Common material is copied to the new project
B. Both Brick materials are copied to the new project
C. Neither Brick materials are copied to the new project
D. A dialog will appear asking you if you want the required Brick material to be copied to the new project

5. To reduce the file size of a project, I should use the following tool:

A. Transfer Project Standards
B. Unload files (to unload any external files)
C. Audit
D. Purge

ANSWERS: 1) A & B; 2) T; 3) T; 4) A; 5) D

About the Author
Autodesk
Certified Instructor

Elise Moss has worked for the past thirty years as a mechanical designer in Silicon Valley, primarily creating sheet metal designs. She has written articles for Autodesk's Toplines magazine, AUGI's PaperSpace, DigitalCAD.com and Tenlinks.com. She is President of Moss Designs, creating custom applications and designs for corporate clients. She has taught CAD classes at Laney College, DeAnza College, Silicon Valley College, and for Autodesk resellers. Autodesk has named her as a Faculty of Distinction for the curriculum she has developed for Autodesk products and she is a Certified Autodesk Instructor. She holds a baccalaureate degree in mechanical engineering from San Jose State.

She is married with three sons. Her older son, Benjamin, is an electrical engineer. Her middle son, Daniel, works with AutoCAD Architecture in the construction industry. Her youngest son works as a warehouse supervisor. Her husband, Ari, has a distinguished career in software development.

Elise is a third-generation engineer. Her father, Robert Moss, was a metallurgical engineer in the aerospace industry. Her grandfather, Solomon Kupperman, was a civil engineer for the City of Chicago.

She can be contacted via email at elise_moss@mossdesigns.com.

More information about the author and her work can be found on her website at www.mossdesigns.com.

Other books by Elise Moss

AutoCAD 2022 Fundamentals
Revit 2022 Certification Guide

Notes: